# *STUDENT SOLUTIONS MANUAL*

## RADIATION
## DETECTION
## And
## MEASUREMENT
### *Fourth Edition*

by
### Glen F. Knoll
*University of Michigan*

### The Student Solutions Manual was authored by
### David K. Wehe
*University of Michigan*

**WILEY**

John Wiley & Sons, Inc.

Founded in 1807, John Wiley & Sons, Inc. has been a valued source of knowledge and understanding for more than 200 years, helping people around the world meet their needs and fulfill their aspirations. Our company is built on a foundation of principles that include responsibility to the communities we serve and where we live and work. In 2008, we launched a Corporate Citizenship Initiative, a global effort to address the environmental, social, economic, and ethical challenges we face in our business. Among the issues we are addressing are carbon impact, paper specifications and procurement, ethical conduct within our business and among our vendors, and community and charitable support. For more information, please visit our website: www.wiley.com/go/citizenship.

ISBN-13     978-0-470-64972-5


10  9  8  7  6  5  4  3  2  1

Printed and bound by Lightning Source

# Table of Contents

# Student Solutions Manual for
# *Radiation Detection and Measurements*
# (4th Edition)

This *Student Solutions Manual* contains detailed Solutions for the odd-numbered problems in the text *Radiation Detection and Measurement*. These Solutions are intended for student use and study while using the textbook. The even-numbered Solutions are excluded from this *Student Solutions Manual*, so that that they will not be available to students, and instructors may use them for student assessment.

Origin of this *Student Solutions Manual*

As a way to learn *Mathematica* while teaching the associated laboratory course, this author gradually compiled a complete set of solutions to the *Radiation Detection and Measurement* problems. Unfortunately, while many students appreciated having the solutions, and in particular the detailed manner in which they are derived, they did not appreciate the elegant but unfamiliar *Mathematica* language needed to produce them. Of course, the intent was to teach them an elegant software tool for doing mathematics while providing them with solutions that they could interact with. Perhaps this was a bit unrealistic to expect (particularly at the undergraduate level) since they were struggling just to absorb the course material and labs. So, at the behest of the publisher, and with the help of a very patient and bright student (Adam Schutte), these have been translated into plain English for your edification, and published in the form of this *Student Solutions Manual*.

There are, undoubtedly, some errors contained within, and for these, the author apologizes in advance. These will hopefully be reported and corrected in future printings.

D. K. Wehe
Nuclear Engineering and Radiological Sciences
University of Michigan
April 1, 2011

---

# Radiation Sources

■ **Problem 1.1. Radiation Energy Spectra: Line vs. Continuous**

Line (or discrete energy): a, c, d, e, f, and i.
Continuous energy: b, g, and h.

■ **Problem 1.3. Nuclear decay and predicted energies.**

We write the conservation of energy and momentum equations and solve them for the energy of the alpha particle. Momentum is given the symbol "p", and energy is "E". For the subscripts, "al" stands for alpha, while "b" denotes the daughter nucleus.

$$p_{al} + p_b = 0 \qquad \frac{p_{al}^2}{2\,m_{al}} = E_{al} \qquad \frac{p_b^2}{2\,m_b} = E_b \qquad E_{al} + E_b = Q \quad \text{and} \quad Q = 5.5\,\text{MeV}$$

Solving our system of equations for $E_{al}$, $E_b$, $p_{al}$, $p_b$, we get the solutions shown below. Note that we have two possible sets of solutions (this does not effect the final result).

$$E_b = 5.5\left(1 - \frac{m_{al}}{m_{al} + m_b}\right) \qquad E_{al} = \frac{5.5\,m_{al}}{m_{al} + m_b}$$

$$p_{al} = \mp \frac{3.31662\,\sqrt{m_{al}}\,\sqrt{m_b}}{\sqrt{m_{al} + m_b}} \qquad p_b = \pm \frac{3.31662\,\sqrt{m_{al}}\,\sqrt{m_b}}{\sqrt{m_{al} + m_b}}$$

We are interested in finding the energy of the alpha particle in this problem, and since we know the mass of the alpha particle and the daughter nucleus, the result is easily found. By substituting our known values of $m_{al} = 4$ and $m_b = 206$ into our derived $E_{al}$ equation we get:

$$E_{al} = 5.395\,\text{MeV}$$

Note : We can obtain solutions for all the variables by substituting $m_b = 206$ and $m_{al} = 4$ into the derived equations above :

$$E_{al} = 5.395\,\text{MeV} \qquad E_b = 0.105\,\text{MeV} \qquad p_{al} = \mp 6.570\,\sqrt{\text{amu} * \text{MeV}} \qquad p_b = \pm 6.570\,\sqrt{\text{amu} * \text{MeV}}$$

■ **Problem 1.5.** $^{235}U$ **Fission Energy Release.**

Using the reaction $^{235}U \rightarrow {}^{117}\text{Sn} + {}^{118}\text{Sn}$, and mass values, we calculate the mass defect of:

$$M\left(^{235}U\right) - \left[M(^{117}\text{Sn}) + M(^{118}\text{Sn})\right] = \Delta M$$

and an expected energy release of $\Delta M c^2$.

$$Q = (235.0439 - (116.9029 + 117.9016))\,\text{AMU} \times \frac{931.5\,\text{MeV}}{\text{AMU}} = 223\,\text{MeV}$$

This is one of the most exothermic reactions available to us. This is one reason why, of course, nuclear power from uranium fission is so attractive.

### Problem 1.7. Accelerated particle energy.

The energy of a particle with charge q falling through a potential $\Delta V$ is $q\Delta V$. Since $\Delta V = 3$ MV is our maximum potential difference, the maximum energy of an alpha particle here is $q*(3$ MV$)$, where q is the charge of the alpha particle (+2). The maximum alpha particle energy expressed in MeV is thus:

> **Energy = 3 Mega Volts × 2 Electron Charges = 6. MeV**

### ■ Problem 1.9. Neutron energy from D-T reaction by 150 keV deuterons.

We write down the conservation of energy and momentum equations, and solve them for the desired energies by eliminating the momenta. In this solution, "a" represents the alpha particle, "n" represents the neutron, and "d" represents the deuteron (and, as before, "p" represents momentum, "E" represents energy, and "Q" represents the Q-value of the reaction).

$$p_a + p_n = p_d \qquad \frac{p_a^2}{2\,m_a} = E_a \qquad \frac{p_n^2}{2\,m_n} = E_n \qquad \frac{p_d^2}{2\,m_d} = E_d \qquad E_a + E_n = E_d + Q$$

Next we want to solve the above equations for the unknown energies by eliminating the momenta. (Note : Using computer software such as Mathematica is helpful for painlessly solving these equations).

We evaluate the solution by plugging in the values for particle masses (we use approximate values of "$m_a$," "$m_n$,"and "$m_d$" in AMU, which is okay because we are interested in obtaining an energy value at the end). We define all energies in units of MeV, namely the Q-value, and the given energy of the deuteron (both energy values are in MeV). So we substitute $m_a = 4$, $m_n = 1$, $m_d = 2$, $Q = 17.6$, $E_d = 0.15$ into our momenta independent equations. This yields two possible sets of solutions for the energies (in MeV). One corresponds to the neutron moving in the forward direction, which is of interest.

$$E_n = 13.340 \text{ MeV} \qquad E_a = 4.410 \text{ MeV}$$
$$E_n = 14.988 \text{ MeV} \qquad E_a = 2.762 \text{ MeV}$$

Next we solve for the momenta by eliminating the energies. When we substitute $m_a = 4$, $m_n = 1$, $m_d = 2$, $Q = 17.6$, $E_d = 0.15$ into these equations we get the following results.

$$p_n = \frac{p_d}{5} \mp \frac{1}{5}\sqrt{2}\,\sqrt{3\,p_d^2 + 352} \qquad p_a = \frac{1}{10}\left(8\,p_d \pm 2\sqrt{2}\,\sqrt{3\,p_d^2 + 352}\right)$$

We do know the initial momentum of the deuteron, however, since we know its energy. We can further evaluate our solutions for $p_n$ and $p_a$ by substituting:

$$p_d = \sqrt{2 \times 2 \times 0.15}$$

The particle momenta ( in units of $\sqrt{\text{amu} * \text{MeV}}$ ) for each set of solutions is thus:

$$p_n = -5.165 \qquad p_a = 5.940$$
$$p_n = 5.475 \qquad p_a = -4.700$$

The largest neutron momentum occurs in the forward (+) direction, so the highest neutron energy of 14.98 MeV corresponds to this direction.

## Radiation Interaction Problems

- **Problem 2.1 Stopping time in silicon and hydrogen.**

Here, we apply Equation 2.3 from the text.

$$T_{stop} = \frac{1.2 \, \text{range} \sqrt{\frac{\text{mass}}{\text{energy}}}}{10^7}$$

Now we evaluate our equation for an alpha particle stopped in silicon. We obtained the value for "range" from Figure 2.8 (converting from mass thickness to distance in meters by dividing by the density of Si $\simeq 2330 \, \text{mg/cm}^3$). The value for "mass" is approximated as 4 AMU for the alpha particle, and the value for "energy" is 5 MeV. We substitute range $= 22 \times 10^{-6}$, mass $= 4$ and energy $= 5$ into Equation 2.3 to get the approximate alpha stopping time (in seconds) in silicon.

$$T_{stop} = 2.361 \times 10^{-12} \text{ seconds}$$

Now we do the same for the same alpha particle stopped in hydrogen gas. Again, we obtain the value for "range" (in meters) from Figure 2.8 in the same manner as before (density of H $\simeq .08988 \, \text{mg/cm}^3$), and, of course, the values for "mass" and "energy" are the same as before (nothing about the alpha particle has changed). We substitute range $= 0.1$, mass $= 4$ and energy $=5$ into Equation 2.3 to get the approximate alpha stopping time (in seconds) in hydrogen gas.

$$T_{stop} = 1.073 \times 10^{-8} \text{ seconds}$$

The results from this problem tell us that the stopping times for alphas range from about picoseconds in solids to nanoseconds in a gas.

- **Problem 2.3. Energy loss of 1 MeV alpha in 5 microns Au.**

From Figure 2.10, we find that $\frac{-1}{\rho} \frac{dE}{dx} \simeq 380 \frac{\text{MeV*cm}^2}{g}$. Therefore, $\frac{dE}{dx} \simeq 380 \frac{\text{MeV*cm}^2}{g} * \rho$ (ignoring the negative sign will not affect the result of this problem).

$$\text{Energy loss} = \frac{\rho \, (dE/dx) \, \Delta x}{\rho}$$

We substitute $dE/dx = \frac{380 \, \text{MeV cm}^2 \, \rho}{\text{gram}}$, $\rho = \frac{19.32 \, \text{grams}}{\text{cm}^3}$ and $\Delta x = 5$ microns to get the energy loss of the 1 MeV $\alpha$-particle in 5 $\mu$m Au (in non-SI units).

$$\text{Energy loss} = \frac{36\,708 \text{ MeV microns}}{\text{cm}}$$

The result in SI units is thus:

$$\text{Energy loss} = 3.671 \times 10^6 \text{ eV}$$

Since this energy loss is greater than the initial energy of the particle, all of the $\alpha$-particle energy is lost before 5 $\mu m$. *Note the small range of the* $\alpha$ *, i.e. ~ $\mu m$ per MeV.*

### ■ Problem 2.5. Compton scattering.

This problem asks for the energy of the scattered photon from a 1 MeV photon that scattered through 90 degrees. We use the Compton scattering formula (Equation 2.17). We write the Compton scattering formula, defining the scattering angle ("$\theta$") as 90 degrees and the photon energy ("$E_0$") as 1 MeV.

$$\text{Energy} = \frac{E_0}{\frac{(1-\text{Cos}[\theta])\,E_0}{m_e * c^2} + 1}$$

We substitute $\theta = 90°$ and $E_0 = 1$ MeV to get the energy of the scattered photon in MeV.

$$\text{Energy} = 0.338 \text{ MeV}$$

### ■ Problem 2.7. The dominant gamma ray interaction mechanism.

See Figure 2-20 and read off the answers (using the given gamma-ray energies and the Z-number for the given absorber in each part).
Compton scattering: a, b, and d.
Photoelectric absorption: c
Pair production: e

### ■ Problem 2.9. Definitions

See text.

### ■ Problem 2.11. 1 J of energy from 5 MeV depositions.

We are looking for the number of 5 MeV alpha particles that would be required to deposit 1 J of energy, which is the same as looking for how many 5 MeV energy depositions equal 1 J of energy. We expect the number to be large since 1 J is a macroscopic unit of energy. To find this number, we simply take the ratio between 1 J and 5 MeV, noting that 1 MeV = 1.6 x $10^{-13}$ $J$.

The number of 5 MeV alpha particles required to deposit 1 J of energy is thus:

$$n = \frac{1 \text{ Joule}}{5 \text{ MeV}} = 1.25 \times 10^{12} \text{ alpha particles}$$

### ■ Problem 2.13. Exposure rate 5 m from 1 Ci of $^{60}$ Co.

We use the following equation for exposure rate:
Exposure Rate $= \Gamma_s \frac{\alpha}{d^2}$   (Equation 2.31)

where $\alpha$ is the source activity, d is the distance away from the source, and $\Gamma_s$ is the exposure rate constant. The exposure rate constant for Co-60 is 13.2 $R - \text{cm}^2/\text{hr} - \text{mCi}$ (from Table 2.1). The exposure rate 5 m from a 1 Ci Co-60 source is thus:

$$\text{Exposure Rate} = \Gamma_s \frac{\alpha}{d^2} \quad where\ we\ substitute\ \alpha = 1\ \text{Ci},\ d = 5\ m,\ \Gamma_s = 13.2\ \frac{R - cm^2}{hr - mCi}$$

$$\text{Exposure Rate} = \frac{0.528\ cm^2\ R}{hr\ mm^2} = \frac{52.8\ mR}{hr}$$

The result in SI units :

$$\text{Exposure Rate} = \frac{3.78 \times 10^{-9}\ C}{kg - s}$$

## ■ Problem 2.15. Fluence-dose calculations for fast neutron source.

The Cf source emits fast neutrons with the spectrum N(E) dE given in the text by Eqn. 1.6. Each of those neutrons carries a dose h(E) that depends on its energy as shown in Fig. 2.22(b). To get the total dose, we have to integrate N(E)h(E) over the energy range.

First, let's consider only the neutron dose h(E). In the MeV range, we need a linear fit to the log h-log E plot of Figure 2.22(b) by using two values read off the curve:

$$\text{Log}_{10}\left(10^{-12}\right) = b + m\,\text{Log}_{10}(0.01)$$
$$\text{Log}_{10}\left(10^{-10}\right) = b + m\,\text{Log}_{10}(1.0)$$

Solving the system of equations above yields :

$$m = 1 \quad and \quad b = -10$$

This result gives us an approximate functional form for h(E) $\left[\text{in Sv} - cm^2\right] = 10^{-10}\ E\ \text{[MeV]}$. Recall 1 Sv = 100 Rem = $10^5$ mrem. In a moment, we'll integrate N(E)h(E) to get the total dose-area, but first check the normalization of N(E):

$$\text{norm} = \int_0^\infty \sqrt{E}\ e^{-\frac{E}{1.3}}\,dE = 1.31359$$

We'll need this normalization factor because we will want to use N(E)/norm as the probability that a source neutron has energy E. Now, a source neutron $- cm^2$ arriving at the person delivers a dose (in Sv) of:

$$\text{neutron} - \text{dose} = \frac{\int_0^\infty E^{3/2}\ e^{-\frac{E}{1.3}}\,dE}{10^{10}\ \text{norm}} = 1.95 \times 10^{-10}\ \text{Sv}$$

We now need to multiply this by the number of neutrons produced by 3 micrograms of Cf-252 (2.3 x $10^6$n/sec-mg) at 5 meters over 8 hours:

$$\text{dose} - \text{equivalent} = \frac{2.3 \times 10^6\ (3\ \mu g)\ (8 \times 3600\ \text{sec})}{\mu\,g\ \text{sec}\left(4\,\pi\,500^2\right)} = 63\,254.5\ \text{neutrons}/cm^2$$

So the total dose equivalent is given by (using $10^5$mrem/Sv):

$$\text{dose equivalent}_{\text{neutrons}} = 1.23\ \text{mrem}$$

This is a small dose, comparable to natural background.

Aside: What about the dose from the gamma rays? They are high-energy gammas and the source emits 9.7 gammas per fission. Using a value of $h_E \sim 5 \times 10^{-12}$ Sv $-$ cm$^2$:

$$\text{dose equivalent}_\gamma = \frac{\left(2.3 \times 10^6 \text{ neutrons}\right) (3 \, \mu\text{g}) \, (8 \times 3600 \text{ sec}) \, (5 \text{ Sv cm}^2) \, (100 \text{ Rad}) \, (1000 \text{ mRad})}{(1 \, \mu\text{g sec}) \left(10^{12} \text{ neutrons}\right) (4 \, \pi \, 500 \text{ cm}^2) \, (\text{Sv Rad})} = 0.0316 \text{ mRad}$$

So the gamma dose is even smaller than that from the fast neutrons, as expected. Fast neutrons have a high quality factor, i.e., they produce a heavy charged particle when they interact, and therefore do a lot more biological damage than the light electrons produced when gamma rays interact in materials.

---

# Counting statistics problems

- ## Problem 3.1. Effect of increasing number of trials on sample variance.

The relative variance of the variable x, i.e., $\frac{\sigma_x^2}{<x>} = \frac{<x^2>}{<x>^2} - 1$, is dependent only on the ratio of the means of $x^2$ and x. It does not depend upon the uncertainty in those quantities. Since these means are not expected to change with more samples, the relative variance (i.e., 2% of the mean) shouldn't change. Note that this conclusion is independent of the type of distribution (Poisson, Guassian, Binomial, etc.) for x. However, for any quantity that is derived from measurements, such as the mean <x>, the uncertainty in that quantity improves with additional measurements as shown by: $\sigma_{<x>} = \sqrt{\frac{<x>}{N}}$.

- ## Problem 3.3. Statistics of males occuring in random population samples.

The mean is well known to be (prob of success)*(number of trials) =0.75 N  The probability of success of any one trial (drawing a male) is large (so Poisson statistics is not valid) and the sample size is only 15. Binomial statistics therefore apply. We substitute n = 15 and p = 0.75 in the equations below to find the mean $(\bar{x})$, variance $(\sigma^2)$ and standard deviation $(\sigma)$.

$$\bar{x} = n \times p = 11.25$$

$$\sigma^2 = n\,p\,(1-p) = 2.81$$

$$\sigma = \sqrt{\sigma^2} = \sqrt{n\,p\,(1-p)} = 1.68$$

- ## Problem 3.5. Statistics of errors in computer program statements.

Poisson statistics applies to this problem because the probability of success (an error) is low, but the expected number of successes is not >~20 (the expected number of successes is 250/60 ≈ 4.17, so we cannot apply Gaussian statistics).

a) Here, we simply define the expected mean and standard deviation of a Poisson distribution with an expected number of successes $(\bar{x}$ = np) of 250/60. Of course, the expected number of successes is the same as the expected mean (they are both equal to np), and in Poisson statistics, the standard deviation is the same as the square root of the expected mean. This is reflected in the results shown below:

$$\bar{x} = \frac{250}{60} = 4.17$$

$$\sigma = \sqrt{\bar{x}} = 2.04$$

b) Next, we define a Poisson distribution with $\bar{x}$= 100/60 (expected number of successes), and evaluate the Poisson probability distribution function at k=0 successes. The probability that a 100-statement program will be free of errors is thus:

$$\text{Probability} = \frac{e^{-\bar{x}}\,\bar{x}^k}{k!} = 0.189$$

### Problem 3.7. Source + Background -> Net counts and uncertainty

We are asked to find net = (S+B) - B and $\sigma_{net}$. Finding $\sigma_{net}$ is a straight forward error propagation since all counts are taken for 1 minute. Below we define the equation for the net counts and the standard error propagation formula (Eqn. 3.37). As a shorthand notation, the error propagation equation is expressed as a dot product between the two vectors representing the squared partial derivatives and the corresponding variances. The variable "sb" refers to (S+B) and "b" refers to B in the equation for "net".

$$net = sb - b$$

$$\sigma_{net} = \sqrt{\left\{ \left( \frac{\partial\, net}{\partial\, sb} \right)^2, \left( \frac{\partial\, net}{\partial\, b} \right)^2 \right\} \cdot \{\sigma_{sb}{}^2, \sigma_b{}^2\}}$$

We substitute sb=561, b=410, $\sigma_{sb}{}^2 = sb$ and $\sigma_b{}^2 = b$ to get the net number of counts and the expected uncertainty in the net number of counts ($\sigma_{net}$).

> **net = 151 counts**

> $\sigma_{net} = 31.16$

### ■ Problem 3.9. Results using optimal counting times for Problem 3.8.

We first solve the equation giving the optimal division of time (Eq. 3.54). This is done below, where we have again defined the variables "sb" to denote the total counts of (S+B), and "b" to denote the background B in the equation. First we solve the system of equations for $T_{sb}$ and $T_b$, and then we substitute the measured values for the count rates (84.6 and 7.3 counts/min, respectively) and the total amount of time allowed for the measurements to be done (20 min) to find the numerical values of $T_{sb}$ and $T_b$.

$$T_{sb} = \sqrt{\frac{sb}{b}}\; T_b \qquad T_b + T_{sb} = T_{tot}$$

The resulting solutions for $T_{sb}$ and $T_b$ solved from the equations above is thus:.

$$T_{sb} = \frac{\sqrt{\frac{sb}{b}}\; T_{tot}}{\sqrt{\frac{sb}{b}} + 1} \qquad T_b = \frac{T_{tot}}{\sqrt{\frac{sb}{b}} + 1}$$

We substitute sb = 84.6, b = 7.3 and $T_{tot} = 20$ to get the numerical values of $T_{sb}$ and $T_b$ in minutes (i.e. the optimal times for measuring (S+B) and B, respectively).

> $T_{sb} = 15.46$ minutes     $T_b = 4.54$ minutes

We now calculate the uncertainty in the net count rate using error propagation. When defining the net count rate, we denote the number of counts over the new optimal time intervals as "$n_{sb}$" and "$n_b$," respectively, and the optimal counting times as "$t_{sb}$" and "$t_b$," respectively. We then use the basic formula for error propagation with the appropriate variables (as in problem 3.7, the dot between the two bracketed quantities signifies the dot product between the two, just as if we thought of them as two vectors). Next we substitute in known values to get the expected error in the net count rate.

$$\text{net} = \frac{n_{sb}}{t_{sb}} - \frac{n_b}{t_b}$$

$$\sigma_{\text{net count rate}} = \sqrt{\left\{\left(\frac{\partial \text{net}}{\partial n_{sb}}\right)^2, \left(\frac{\partial \text{net}}{\partial n_b}\right)^2\right\} \cdot \left\{\sigma_{n_{sb}}^2, \sigma_{n_b}^2\right\}}$$

We substitute $\sigma_{n_{sb}}^2 = n_{sb} = 84.6 \, t_{sb}$, $\sigma_{n_b}^2 = n_b = 7.3 \, t_b$, $t_{sb} = 15.5$ and $t_b = 4.5$ to get the expected uncertainty in the net count rate when the optimal time intervals are used.

$$\sigma_{\text{net count rate}} = 2.66$$

The improvement factor is thus 3.03/2.66.

■ **Problem 3.11. Better to increase source or decrease background?**

Recall our relationship that the relative uncertainty (fractional standard deviation squared, or $\epsilon^2$) in the source is given by:

$$\frac{1}{T} \frac{\left(\sqrt{S+B} + \sqrt{B}\right)^2}{S^2} \quad \text{(from Eqn. 3.55)}$$

(a). If the $S \gg B$, then this becomes $\frac{1}{ST}$. Doubling the source strength is the best choice since the background is nearly irrelevant.

(b). If $S \ll B$, then this becomes $\frac{4B}{S^2 T}$. Doubling source improves the ratio by 4 times, whereas halving background only improves ratio by 2 times. Doubling the source strength is again the best choice.

■ **Problem 3.13. Percent standard deviation between the activity ratio of two sources.**

The ratio of the activity of Source B to Source A is simply given by:

$$\text{Ratio} = \frac{[(\text{Source } B + \text{background count rate}) - (\text{background count rate})]}{[(\text{Source } A + \text{background count rate}) - (\text{background count rate})]}$$

This is represented below where we denote the number of counts by "c," measurement time by "t," background by "b," Source B + background by "bb," and Source A + background by "ab."

$$\text{Ratio } R = \frac{\dfrac{c_{bb}}{t_{bb}} - \dfrac{c_b}{t_b}}{\dfrac{c_{ab}}{t_{ab}} - \dfrac{c_b}{t_b}}$$

We substitute $c_{ab} = 251$, $t_{ab} = 5$, $c_{bb} = 717$, $t_{bb} = 2$, $c_b = 51$ and $t_b = 10$ to get the ratio of the activity of Source B to Source A.

**Ratio $R = 7.84$**

Next, we define the explicit error propagation formula with the appropriate variables (as in previous problems, using the dot product notation under the square root) and give the known values in counts and minutes, respectively.

$$\sigma = \sqrt{\left\{\left(\frac{\partial R}{\partial c_{ab}}\right)^2, \left(\frac{\partial R}{\partial c_{bb}}\right)^2, \left(\frac{\partial R}{\partial c_b}\right)^2\right\}.\{c_{ab}, c_{bb}, c_b\}}$$

We substitute $c_{ab} = 251$, $t_{ab} = 5$, $c_{bb} = 717$, $t_{bb} = 2$, $c_b = 51$ and $t_b = 10$ to get the percent standard deviation in the ratio of the activity of Source B to Source A.

$$\sigma_{B/A} = 0.635$$

## ■ Problem 3.15. Uncertainty in groups of measurements.

(a). The data fluctuations are expected to be statistical if the standard deviation of the sample population is the square root of the mean. This seems true, but we check this with a Chi-squared distribution to be sure.

Here, we define the student's data set and calculate the standard deviation of that data set.

$$\text{data} = \{25, 35, 30, 23, 27\}$$

$$\sigma = 4.69$$

Next, we take the square root of the mean (which is 28).

$$\sqrt{\overline{x}} = 5.29$$

Here, we define chi-squared ($\chi^2$) for the data set (using Eqn. 3.36).

$$\chi^2 = \frac{(N-1)\sigma^2}{\overline{x}}$$

We substitute N=5, $\sigma^2$=22 and $\overline{x}$=28 to get the value of $\chi^2$ for our data set.

$$\chi^2 = \frac{22}{7}$$

This is our "measured" value of $\chi^2$. One can also calculate the expected value of $\chi^2$ from data values drawn from a predicted distribution. (The $\chi^2$ distribution is defined as the distribution of the quantity $\sum_{i=1}^{n} x_i^2$, where the $x_i$ are random variables following a normal distribution that has a unit variance and a mean value of zero).

To be consistent with the approach taken in the textbook, we actually want to calculate 1 minus the $\chi^2$ "**C**umulative **D**istribution **F**unction" (CDF). The $\chi^2$ CDF is the integral from zero up to some argument, which in this case is $\frac{22}{7}$. The function CDF(x) gives the probability that the expected value of $\chi^2$ ranges between 0 and the value x, assuming the data follow the normal distribution. The complement of this is then the probability that the expected value of $\chi^2$ is larger than this value, which is what the textbook uses. One can look these values up in statitics tables, but we use *Mathematica* to do this calculation for us:

$$1 - \text{CDF}\left[\chi^2\left(4\,\text{degrees} - \text{of} - \text{freedom}, \frac{22}{7}\right)\right] = 0.534$$

Because this value is close to 0.5, the data are random (i.e. a true Poisson distribution would have a $\chi^2$ probability of 0.5). As a side note, this probability could also have been estimated using the $\chi^2$ distribution table.

b) This question is asking that since the data appears to be random, what is the EXPECTED standard deviation of the MEAN of 5 single measurements using just these data. For this data set, given a mean , $\sigma_{\bar{x}} = \sqrt{\dfrac{\bar{x}}{N}}$ (Eqn. 3.44).

Using the provided data, the expected standard deviation in the mean of the data set is:

$$\sigma_x = \sqrt{\dfrac{\bar{x}}{N}} = 2.37$$

(c). Now, suppose we have the 30 measurements of the mean from the 30 students. What would we expect to measure for the variance of the set of these mean values $\{x_1, x_2, x_3,..., x_{30}\}$?

The variance estimated in (b) above IS the expected fluctuation when samples are drawn identically from that population. So we expect that the 30 mean values WILL show this variance (i.e. we would expect the sample variance in this situation to be the result above squared, or $s^2 = \sigma_{\bar{x}}^2$; this is calculated to be $\approx 5.59$).

(d). Suppose we now average these 30 values to get a better estimate of the mean. What is the standard deviation for the mean?

One way to look at this is to see it as 5 x 30 data points and calculate $\sigma_{\bar{x}}$ . The standard deviation goes down by $\sqrt{30}$ , so the expected standard deviation of the mean when we use all 30 mean values will be 0.432049.

Another way is to calculate the standard deviation of the average of the averages. If we define $<\bar{x}>$ as the average of the 30 students individual means, then:

$$\sigma_{<\bar{x}>} = \sqrt{\dfrac{<\bar{x}>}{N}}$$

which, of course, turns out to be exactly the same formula.

- **Problem 3.17. Uncertainty in the difference between two measurements.**

Suppose we are given that a set of I counts $\{N_i\}$, each taken over a period of time $t_I$ results in an average rate $<r>$. The uncertainty in the average rate can be determined from error propagation once we write the average rate in terms of the measured counts:

$$<r> = (1/I)(N_1/t_I + N_2/t_I + \ldots N_I/t_I) = (1/It_I)(N_1 + N_2 + \ldots N_I) = N_{total}/t_{total}$$

so

$$\sigma_{<r>}{}^2 = (1/It_I)^2 \{N_1 + N_2 + \ldots + N_I) = N_{total}/t_{total}{}^2$$

We know that the total number of counts $N_{total} = <r> t_{total}$. Since for this problem, we are given $<r>$ and $t_{total}$ for each group, we can find $N_{total}$ for each group.

Wtih $N_A$ and $N_B$ ( total counts from Group A and Group B), we want to find out if the difference between $<r>_A$ and $<r>_B$ is significant. Define this difference as:

$$\Delta = \frac{N_A}{T_A} - \frac{N_B}{T_B}.$$

where $T_A$ is the total time for Group A (i.e., I * $t_I$ ), and similarly for Group B.

Using error propagation on these independent measurements,

$$\sigma_\Delta{}^2 = \left(\frac{\sigma_{N_A}}{T_A}\right)^2 + \left(\frac{\sigma_B}{T_B}\right)^2 = (<r_A>/T_A + <r_B>/T_B), \text{ since } \sigma_{N_A}{}^2 = N_A \text{ (i.e., the standard deviation in a number of counts is equal to}$$

the square root of that number).

If both groups are making identical measurements, we expect the probability of observing a given value of $\Delta$, i.e., P($\Delta$), to be

Gaussian with $<\Delta>=0$ and $\sigma_\Delta = \sqrt{<r>\left(\frac{1}{T_A} + \frac{1}{T_B}\right)}$

We look at our measured value of $\Delta$, and see if it lies within $\pm\sigma_\Delta$ of 0. We will look at $\frac{\Delta_{meas}}{\sigma_{\Delta_{theory}}}$ and if this value is $>> 1$, then the difference is significant since the probability of observing our value of $\Delta$ would be small.

Below we define $\frac{\Delta}{\sigma_\Delta}$, and substitute the known values for "$N_A$," "$N_B$," "$T_A$," and "$T_B$"

$$\frac{\Delta}{\sigma_\Delta} = \frac{\frac{N_A}{T_A} - \frac{N_B}{T_B}}{\sqrt{\frac{N_A}{T_A^2} + \frac{N_B}{T_B^2}}}$$

We substitute $N_A = 2162.4 \times 10$, $N_B = 2081.5 \times 20$, $T_A = 10$ and $T_B = 20$ to get the value of $\frac{\Delta}{\sigma_\Delta}$.

$$\frac{\Delta}{\sigma_\Delta} = 4.52$$

Since the difference between the two measured averages is 4.5 standard deviations from the expected mean (of zero), the difference is significant (i.e. there is a very small probability of observing such a value). We conclude that the premise that both groups were making the same measurement is highly unlikely.

## Problem 3.19. Measuring half-life and its expected standard deviation of a source.

a) To solve this, we write the expression for the half life from the data measured, two measured counts taken over two measurement times that are separated by time t. The background is assumed to have no uncertainty. This expression is shown below, where the time separation between the measurements is denoted by "t," number of measured counts by $n_1$ and $n_2$, the measurement times by $t_1$ and $t_2$. We denote the background count rate by $b_{rate}$, and then the calculated half-life $t_{half}$ is given by:

$$t_{half} = -\frac{\ln(2)\,t}{\ln\left(\frac{\frac{n_2}{t_2} - b_{rate}}{\frac{n_1}{t_1} - b_{rate}}\right)}$$

We substitute $t = 24$ hours, $n_2 = 914$, $t_2 = 10$ minutes, $n_1 = 1683$, $t_1 = 10$ minutes and $b_{rate} = 50$ minute$^{-1}$ to get the calculated half-life of the source:

$t_{half} = 15.8$ hours

b) Now to determine the uncertainty in this value, we apply the error propagation formula. This is shown below (as in previous problems, we use the shorthand dot product notation here for the error propagation formula). Note that we are applying the fact that $\sigma_{n^2} = n$ (which is acceptable because "n" is a number of counts).

$$\sigma_{t_{half}} = \sqrt{\left\{\left(\frac{\partial t_{half}}{\partial n_1}\right)^2, \left(\frac{\partial t_{half}}{\partial n_2}\right)^2\right\}.\{n_1, n_2\}} = \sqrt{\frac{n_1\,t^2\ln^2(2)}{\left(\frac{n_1}{t_1} - b_{rate}\right)^2 t_1{}^2 \ln^4\left(\frac{\frac{n_2}{t_2} - b_{rate}}{\frac{n_1}{t_1} - b_{rate}}\right)} + \frac{n_2\,t^2\ln^2(2)}{\left(\frac{n_2}{t_2} - b_{rate}\right)^2 t_2{}^2 \ln^4\left(\frac{\frac{n_2}{t_2} - b_{rate}}{\frac{n_1}{t_1} - b_{rate}}\right)}}$$

We substitute $t = 24$ hours, $n_2 = 914$, $t_2 = 10$ minutes, $n_1 = 1683$, $t_1 = 10$ minutes and $b_{rate} = 50$ minute$^{-1}$ to get the expected standard deviation of the half-life

$\sigma_{t_{half}} = 1.22$ hours

## ▪ Problem 3.21. Uncertainty in decay constant measurement.

In this problem, we wish to adjust the waiting time to minimize the predicted uncertainty in the decay constant from two measurements. We have an approximate value of the decay constant, $\lambda_p$, which will enable us to predict the number of counts expected at any value of the waiting time. Begin by defining the decay constant and its uncertainty in terms of the measured counts, noting that we expect $n(t_1) = n(t_0)\,e^{-\lambda_p\,\Delta t}$.

$$\lambda = \frac{\ln\left(\frac{n_1}{n_0}\right)}{\Delta t}$$

$$\sigma_\lambda = \sqrt{n_0\left(\frac{\partial \lambda}{\partial n_0}\right)^2 + n_1\left(\frac{\partial \lambda}{\partial n_1}\right)^2}$$

We substitute $n_1 = n_0\,e^{-\lambda\,\Delta t}$ to get the variance in the decay constant.

$$\sigma_\lambda = \sqrt{\frac{e^{\lambda_p \Delta t}}{\Delta t^2 n_0} + \frac{1}{\Delta t^2 n_0}}$$

To find the minimum in the variance with respect to the wait time $\Delta t$, we set the derivative equal to zero:

$$\frac{\partial \sigma_\lambda}{\partial \Delta t} = \frac{e^{\lambda_p \Delta t}\left(\lambda_p \Delta t - 2\right) - 2}{2 \Delta t^3 n_0 \sqrt{\frac{e^{\lambda_p \Delta t}+1}{\Delta t^2 n_0}}} = 0$$

We recognize that the numerator must be zero, so we solve for $\Delta t$:

$$e^{\Delta t \lambda_p}\left(\Delta t \lambda_p - 2\right) - 2 = 0$$

This is an equation of the form $x = 2(1 + e^{-x})$. We plot the left-hand-side and the right-hand-side of the equation, and see that they are equal at about $x = \lambda_p \Delta t = 2.2$.

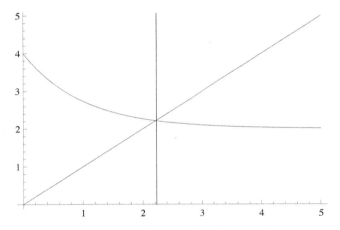

We can also use a nonlinear root finder to find the numeric value for the numerator, which yields a more accurate answer for the optimal wait time $\Delta t$:

$$\Delta t = \frac{2.22}{\lambda_p}$$

■ **Problem 3.23. Calculating fraction of intervals less than $\Delta t$ with a known average count rate.**

The probability of the next event taking place in an interval dt after a delay $\Delta t$ is given by $(r * e^{-r\Delta t})$ dt. (cf. Eqn.3.71). The probability that the time between two events is less than $\Delta t$ is given by the integral of this interval distribution from 0 to $\Delta t$. Of course, this must be the same as the the complement of the probability that no events occur during the time interval $\Delta t$, or $1 - e^{-rt}$. We evaluate this function:

$$\text{Fraction} = 1 - e^{-r\Delta t}$$

We substitute $\Delta t = 10^{-2}$ sec and $r = 100$ sec$^{-1}$ to get the fraction of the intervals that are less than 10 ms when the average event rate is 100 sec$^{-1}$.

$$\text{Fraction} = 0.632$$

## Problem 3.25. Intervals between events.

(a). For a fixed frequency f, the period is T=1/f, and the mean wait time is (1/2)T. For an analogy, think of blindly putting your pencil down on a ruler. The intervals between tick marks is constant. Since all points are equally probable, the average pencil mark is at 1/2 of the scale. Therefore, the average (mean) waiting time until the next bus arrives is (1/2)*(30 minutes) = 15 minutes.

(b). If the interval distribution is not a constant, then the mean wait time is the integral of the probability of arriving during an interval of $\Delta t$ between buses ($P(\Delta t)\, dt$ ), times the mean waiting time for this interval (which is $\Delta t/2$) from part (a) above. The way to realize this is to consider 5 intervals of length 1, and 1 interval of length 5. The probability of an interval of length 1 is 5/6, but the probability of arriving during an interval of length 1 is actually 1/2 (5 of the 10 length scale). So the mean distance (time) to the next tick (bus) for this contrived example would be: 1/2*(1/2) + 1/2*(5/2) = 1.5 .

This concept is expressed below. The probability of an interval t when the mean rate is r is given by Poisson statistics as $e^{-rt}\, r\, dt$. If the average wait for this interval is t/2, then the overall mean wait time is the integral over all possible intervals t. We have noted that the average time between Poisson distributed events (i.e., interval) is the inverse of the rate.

$$\text{average wait time} = \frac{\int_0^\infty \frac{1}{2} r t\, e^{-rt}\, t\, dt}{\int_0^\infty r t\, e^{-rt}\, dt}$$

We substitute $r = \frac{1}{T}$ and evaluate to find the average wait time.

$$\text{average wait time} = T$$

The average waiting time for randomized buses is thus T, twice as long as for buses which are periodic. This makes sense intuitively because for a Poisson process, the average time between events is 1/r = 1/f = T.

## ♀ Supplemental Problem: Professor Kerr's Tricky Problem.

A detector detects gamma rays from a radioactive source with the detected events occurring at a rate such that the average interval between events is a time T. Begin observing at time zero. At what time t will the probability of having detected exactly one event be 0.5?

Start by assuming Poisson statistics are valid. Now we want to look at the probability distribution function. Here, we define our distribution function, "P," as a function of two variables, "n" and "t." We measure time t with respect to T, so our variable "t" actually corresponds to t/T, and "n" is simply the number of counts recorded at time t..

$$P = \frac{t^n}{e^t\, n!}$$

We can plot the probability of observing exactly 1 count at time t by giving "n" that value of 1 and plotting "t" from 0 to 5.

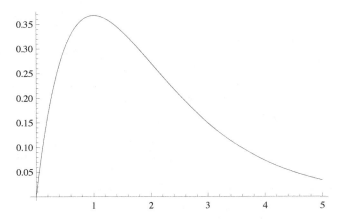

Note that the maximum probability of observing 1 count occurs near t=T ( where t/T=1 on our graph) and takes on a value of only about 0.36. We can find this exact maximum by taking the derivative of the distribution with respect to t, setting it equal to zero, solving for t, and then substituting this value for t back into the distribution.

$$\frac{\partial P}{\partial t} = \frac{e^{-t} n\, t^{n-1}}{n!} - \frac{e^{-t}\, t^{n}}{n!} = 0$$

We solve the above equation for "t" to find the value of "t" for when the probability distribution function reaches its maximum (or when the derivative equals zero).

$$t = 1$$

Now we substitute t=1 and n=1 back into our probability distribution function "P" to get the exact value of the maximum probability of detecting only one event.

$$P = \frac{t^{n}}{e^{t}\, n!} = \frac{1}{e} = 0.368$$

So, to answer the question posed by Professor Kerr, there is no point at which the probability of detecting exactly one event will be 0.5, and, as expected, the maximum probability of this detecting only one event occurs at t=T, and has a value of 1/e. It's a tricky question because the answer is "no time" rather than a specific time.

### ♥ Supplemental Problem: Uncertainty in signal to noise ratio.

The signal to noise ratio, "R," is defined by the net number of counts divided by the noise level (i.e. the statistical uncertainty) of the background. If "T" is the total number of counts from the source, "S," and background, "B," (i.e. S = T-B) then $R = \frac{(T-B)}{\sqrt{B}}$. Find the uncertainty and relative uncertainty in the signal to noise ratio, and evaluate this for S=B.

Solution : First we define the signal to noise ratio. Then we define the uncertainty in R using the error propagation formula (using partial derivative and dot product notation). Then we define the relative uncertainty (fractional uncertainty) in R.

$$R = \frac{T - B}{\sqrt{B}}$$

$$\sigma_R = \sqrt{\left\{\left(\frac{\partial R}{\partial T}\right)^{2}, \left(\frac{\partial R}{\partial B}\right)^{2}\right\}.\{T, B\}} = \frac{1}{2}\sqrt{\frac{B^{2} + 6\,T\,B + T^{2}}{B^{2}}}$$

$$\text{relative uncertainty} = \frac{\sigma_R}{R} = \frac{\sqrt{B}\sqrt{\dfrac{B^2+6TB+T^2}{B^2}}}{2(T-B)}$$

Next we substitute T=2B (since S=B and S=T-B) into the above equations to get R, the uncertainty in R ($\sigma_R$), and the relative uncertainty in R $\left(\frac{\sigma_R}{R}\right)$.

$$R = \sqrt{B} \qquad \sigma_R = \frac{\sqrt{17}}{2} \qquad \text{and} \qquad \frac{\sigma_R}{R} = \frac{\sqrt{17}}{2\sqrt{B}}$$

## ♀ Supplemental problem: Professor Fleming's Attenuation Coefficient and Count Time Statistics

Suppose we have a constant source of monoenergetic particles with strength $I_0$ particles per unit time. We have a detector with constant efficiency and no background. Find the optimum thickness of the sample which minimizes the relative error in the measured value of $\mu$ given that the measurement must be carried out in a fixed measurement time T.

There are two variables at work here: the material thickness (which affects the number of counts obtainable when the sample is in the beam) and the observation times (divided between sample-in $T_s$ and sample-out $T_0$ times, where $T_s + T_0 = T$).

Let $y = \mu t = \ln\left(\frac{I_0}{I_s}\right) = \ln\left(\frac{N_0}{N_s}\right) - \ln\left(\frac{T_0}{T_s}\right)$, where y is the thickness in mfp units.
Then:

$$\sigma_y^2 = \left(\frac{1}{N_0} + \frac{1}{N_s}\right) = \frac{1}{I_0 T_o}\left(1 + e^y \frac{T_0}{T_s}\right) = \frac{1}{I_0 T}\frac{1}{f}\left(1 + e^y \frac{f}{1-f}\right)$$

where $f = \frac{T_0}{T}$. Define the relative variance to be r, where:

$$r = \frac{1}{I_0 T}\frac{1}{fy^2}\left(1 + e^y \frac{f}{1-f}\right)$$

Choosing $I_0 T = 1$ for convenience:

$$r(y, f) = \frac{\frac{f e^y}{1-f} + 1}{y^2 f}$$

Let's plot the shape of the variance surface as a function of sample thickness y and fraction of time spent observing the open beam f:

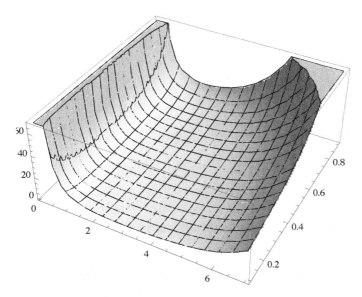

The minimum relative variance occurs at r=3.22404, y=2.55693 and f=0.217811. Note that these can be found by simultaneously solving $\frac{\delta r}{\delta y} = \frac{\delta r}{\delta f} = 0$

- **Supplemental Problem: What is the probability of observing the same (or different by Δ) subsequent number of counts?**

The probability distribution function for the Poisson is given by:

$$\mathbf{pdf} = \frac{e^{-\mu}\,\mu^x}{x!}$$

We calculate, for every observable value of counts x, the probability of observing two subsequent counts that have the same value. So we sum pdf[x, $\mu$]$^2$ for all possible values of x given the mean $\mu$:

$$\mathbf{Probability} = \sum_{x=0}^{\infty}\left(\frac{e^{-\mu}\,\mu^x}{x!}\right)^2$$

A plot of this function is shown below where $\mu$ goes from 0 to 15.

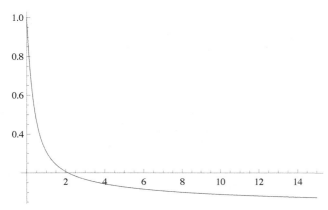

This makes sense. If the mean is small, the probability of two subsequent counts having the same number is higher. Indeed, if the mean of the background is 0, then two subsequent counts have to have the same value (namely, zero!).

18

Suppose we look at the same question using Normal distribution: m is the mean now. In order to use the normal distribution, we must now assume that the mean is large. We integrate the probability of observing two values of x, and integrate over all possible values of x:

$$\text{Probability} = \frac{1}{2\pi m} \int_0^\infty e^{\frac{-(x-m)^2}{m}}\, dx = \frac{\text{erf}\left(\sqrt{m}\right) + 1}{4\sqrt{\pi}\,\sqrt{m}}$$

Below is a plot of y=Erf($\sqrt{m}$) where m goes from 0 to 30.

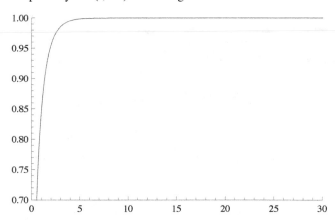

Note that Erf[m] → 1 for m large, so our result becomes $\frac{1}{2\sqrt{\pi m}}$.

Another way to look at this problem is to calculate the probability of observing 0 net counts. That is the same as assuming two subsequent counts that have the same value. We assume two Normal distributions, the difference is a Normal distribution with mean of 0 and standard deviation of $\sqrt{2m}$. We could use this same approach to calculate that two subsequent values differ by any amount, not just 0.

$$\text{Probability} = \frac{e^{-\frac{(x-\mu)^2}{2\sigma^2}}}{\sqrt{2\pi}\,\sigma}$$

We substitute x=0, $\mu$=0 and $\sigma=\sqrt{2m}$ to find the probability as thus:

$$\text{Probability} = \frac{1}{2\sqrt{\pi}\,\sqrt{m}}$$

We, of course, get the same answer, but with a much simpler approach.

### ■ A related problem

We ask the question what is the probability of observing a difference of delta between two measurements from a single distribution with mean $n_{1\,\text{av}}$.

$$\text{Probability} = \frac{1}{2\pi n_{1\,\text{av}}} \int_0^\infty \int_0^\infty \frac{\delta(-\delta - n_1 + n_2)}{e^{\frac{(n_1 - n_{1\,\text{av}})^2}{2n_{1\,\text{av}}}}\, e^{\frac{(n_2 - n_{1\,\text{av}})^2}{2n_{1\,\text{av}}}}}\, dn_1\, dn_2$$

$$\text{Probability} = \frac{e^{-\frac{\delta^2}{4 n_{av}}} \left( \theta(-\delta) \left( \text{erf}\left( \frac{\delta - 2 n_{av}}{2 \sqrt{n_{av}}} \right) + \text{erf}\left( \frac{2 n_{av} + \delta}{2 \sqrt{n_{av}}} \right) \right) + \text{erfc}\left( \frac{\delta - 2 n_{av}}{2 \sqrt{n_{av}}} \right) \right)}{4 \sqrt{\pi} \sqrt{n_{av}}}$$

Note that the answer is a Gaussian distribution with mean of zero, as before, with a new variance $\sqrt{2}$ larger. We now substitute $\delta = 0$ to get the probability.

$$\text{Probability} = \frac{\text{erf}\left( \sqrt{n_{av}} \right) + 1}{4 \sqrt{\pi} \sqrt{n_{av}}}$$

A plot of $y = \text{Erf}\left( \sqrt{x} \right)$ is shown below where x goes from 0 to 100.

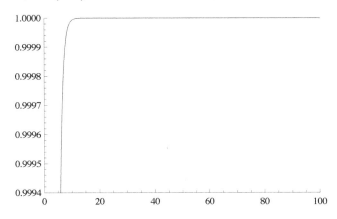

which is the same answer we had before (the problem above has $\delta = 0$ too) if we let Erf[...] -> 1.

---

## General Properties of Radiation Detectors problems

■ **Problem 4.1. Voltage from collected charge Q on capacitance C.**

Since we know that $10^6$ electrons are collected, and we know that the charge of an electron is approximately 1.6 x $10^{-19}$ C, the total charge collected is easily computed. We also know that the amplitude of the voltage pulse is $V = \frac{Q}{C}$. The following expression defines the equation for the voltage, giving the appropriate value for Q in terms of the number of electrons collected and the charge of a single electron and giving the known capacitance:

$$V = \frac{Q}{C}$$

We substitute $Q = \frac{10^6 \text{ electrons } 1.6 \text{ Coulombs}}{10^{19} \text{ electrons}}$ and $C = \frac{100 \text{ Farads}}{10^{12}}$ to get the amplitude of the voltage pulse.

$$V = 0.0016 \text{ Volts}$$

■ **Problem 4.3. Derive expression for mean squared fluctuation in voltage.**

Combine [Eq'n 4.2] $I_0 = r\,Q$ with [Eq'n 4.7] $\overline{\sigma_I^2(t)} = \frac{I_0^2}{rT}$ to solve for $\overline{\sigma_I^2(t)}$ by eliminating $I_0$:

$$I_0 = r\,Q,$$

$$\overline{\sigma_I^2(t)} = \frac{I_0^2}{r\,T}$$

Substituting for $I_0^2$ in the latter expression yields the answer we seek:

$$\overline{\sigma_I^2[t]} = \frac{Q^2\,r}{T}$$

■ **Problem 4.5. DPHS -> IPHS and Counting Curve**

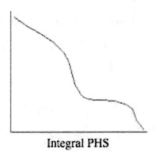

Integral PHS

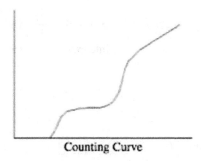

Counting Curve

21

■ **Problem 4.7. Energy resolution.**

The two peaks are separated by 55 keV. We want the peaks separated by at least one FWHM (see the discussion in the textbook). First, we calculate the energy resolution of the 435 keV peak. This is done using R = $\frac{\text{FWHM}}{H_0}$ = $\frac{55\,\text{keV}}{435\,\text{keV}}$. The energy resolution for the 435 keV peak assuming a FWHM of 55 keV is thus:

$$R_{435} = \frac{55}{435} = .126437 = 12.6\,\%$$

Next, we do the same calculation for the 490 keV peak. The energy resolution for the 490 keV peak assuming a FWHM of 55 keV is thus:

$$R_{490} = \frac{55}{490} = .112245 = 11.2\,\%$$

In order for the peaks to be separated by at least 1 FWHM, we take the smaller energy resolution as the value of the system energy resolution. Therefore, the required system energy resolution is about 11.2 % (or less).

■ **Problem 4.9. Electronic noise added to intrinsic resolution.**

The key idea here is that electronic noise adds in quadrature with intrinsic resolution to determine the system's overall resolution, i.e. $\Delta E_{\text{tot}}^2 = \Delta E_{\text{int}}^2 + \Delta E_{\text{noise}}^2$

Using this relationship, we want to evaluate $\Delta E_{\text{tot}}$. This is done below, where we solve for $\Delta E_{\text{tot}}$ and give the appropriate energy resolution values (in fractions, not percents).

$$\Delta E_{\text{tot}} = \sqrt{\Delta E_{\text{int}}^2 + \Delta E_{\text{noise}}^2}$$

We substitute $\Delta E_{\text{int}} = 0.04$, $\Delta E_{\text{noise}} = 0.02$ to get the expected overall pulse height resolution (expressed as a fraction, not a percent).

$$\Delta E_{\text{tot}} = 0.0447$$

■ **Problem 4.11. Solid Angle.**

We use the integral over $\frac{d\Omega}{4\pi} = \frac{\sin\theta\, d\theta\, d\phi}{4\pi}$ for $\phi \in \{0, 2\pi\}$, $\theta \in \{0, 0.25°\}$ to find $\frac{\Omega}{4\pi}$, which is the probability that the laser beam will strike the moon. This integral is expressed below (when $d\phi$ is integrated from 0 to $2\pi$, we get a factor of $2\pi$ in the overall integral).

$$\text{Probability} = \int_0^\theta \frac{2\pi\,(\sin\theta)}{4\pi}\, d\theta$$

The equation for $\frac{\Omega}{4\pi}$ is dependant on the upper limit of $\theta$.

$$\text{Probability} = \frac{1}{2} - \frac{\cos\theta}{2}$$

Next, we substitute $\theta = 0.25°$ to get the probability that the laser will strike the moon.

**Probability = $4.75964 \times 10^{-6}$**

Just for fun, we perform the Taylor series expansion for $Cos[\theta \sim small]$ up to order 2 on our equation for $\frac{\Omega}{4\pi}$.

The Taylor series expansion of the equation for $\frac{\Omega}{4\pi}$ is thus:

**Probability $= \frac{\theta^2}{4}$**

Again, we substitute $\theta = 0.25°$ to get an approximation of the probability that the laser will strike the moon.

**Probability $= 4.75965 \times 10^{-6}$**

This is obviously a good approximation and doesn't require looking up the value of the cosine.

We could also use the approximate expression for $\Omega$ where $d \gg a$, $\frac{\Omega}{4\pi} = \frac{\pi a^2}{4\pi d^2}$ and since $\tan[\theta]=a/d$, $\frac{\Omega}{4\pi} = \frac{\tan^2\theta}{4}$ This is expressed below, where we give the value for $\theta$.

**Probability $= 0.25 \tan^2(0.25°)$**

Another approximation of the probability that the laser will strike the moon is thus:

**Probability $= 4.75971 \times 10^{-6}$**

Again, this is a good approximation.

- ## Problem 4.13. Dead time models for decaying source.

To solve this, we find the analytic solution to the two governing equations for the paralyzable model, the most appropriate model for the GM tube. We take the natural log of both sides of the paralyzable model expression.

$$\ln(m_0) = \ln(n_0) - n_0\,\tau$$

In the next line, we express Eqn. 4.36 (we make the distinction between "$m_0$" and "$m_1$" because one is a measured count rate at time 0 and one is measured count rate after an elapsed amount of time "t").

$$m_1 \ln = -n_0\,\tau\,e^{t(-\lambda)} + n_0 \ln - t\,\lambda$$

Next, we solve our equations for $\ln(n_0)$ to give the analytical solution:

$$\ln(n_0) = \frac{t\,e^{t\lambda}\,\lambda - \ln(m_0) + e^{t\lambda}\ln(m_1)}{-1 + e^{t\lambda}}$$

A simplified form of $\ln(n_0)$ is thus:

$$\ln(n_0) = \frac{e^{t\lambda}\,(t\lambda + \ln(m_1)) - \ln(m_0)}{-1 + e^{t\lambda}}$$

Using this solution, we eliminate the ln on the left hand side by raising "e" to the power of both sides, thus solving for $n_0$, and giving the appropriate known values.

$$n_0 = e^{\frac{t\lambda - \frac{\ln(m_0)}{e^{t\lambda}} + \ln(m_1)}{1 - e^{-t\lambda}}}$$

We substitute $m_0 = 131\,340$, $m_1 = 93\,384$, $\lambda = \frac{\ln(2)}{54.3}$ and $t = 40$ to get the calculated true interaction rate in the G-M tube at 12:00, $n_0$(in min$^{-1}$) for the paralyzable case.

$$n_0 = 200\,691 \text{ min}^{-1}$$

Now we do the same thing for the non-paralyzable case, solving the appropriate governing equations (Eqn. 4.23 and Eqn. 4.35, slightly rearranged).

$$n_0 - m_0 = n_0\,m_0\,\tau$$
$$n_0\,e^{-t\lambda} - m_1 = n_0\,e^{-t\lambda}\,m_1\,\tau$$

The solution for $n_0$.

$$n_0 = \frac{\left(-1 + e^{\lambda t}\right) m_0\,m_1}{m_0 - m_1}$$

We substitute $m_0 = 131\,340$, $m_1 = 93\,384$, $\lambda = \frac{\log(2)}{54.3}$ and $t = 40$ to get the calculated true interaction rate in the G-M tube at 12:00, $n_0$(in min$^{-1}$) for the non-paralyzable case.

$$n_0 = 215\,307 \text{ min}^{-1}$$

Note that the two results are about 8% different depending on which model we chose to be most representative of our true detection system.

■ **Problem 4.15. Dead time.**

The key to this problem is to recognize that two measured rates are given and the ratio of the true source rates are known (if one source is in place, we have a true source rate of n, but if two identical sources are in place, we have a true source rate of 2n). The value of $\tau$ is desired, and background is negligible. Let's assume a non-paralyzable model, i.e. $n = \frac{m}{1 - m\tau}$. We could just use equation 4.32, but let's solve the two equations directly. The two equations for the different measured rate are shown below, where we solve for "$\tau$" by eliminating "n."

$$n = \frac{10\,000}{1 - 10\,000\,\tau}$$
$$2\,n = \frac{19\,000}{1 - 19\,000\,\tau}$$

The counter dead time in seconds for the non-paralyzable model is thus:

$$\tau = 5.26 \times 10^{-6}$$

24

This is the same result as if equation 4.32 was used, of course. We can do the same approach for the paralyzable model, i.e. m = $ne^{-n\tau}$. The corresponding equations for our problem are shown below.

$$10\,000 = n\,x,$$
$$19\,000 = 2\,n\,x^2$$
$$x = e^{-n\tau}$$

Our solution for $\tau$ in fraction form.

$$\tau = \frac{19\ln\left(\frac{20}{19}\right)}{200\,000}$$

The counter dead time in seconds for the paralyzable model is thus:

$$\tau = 4.87 \times 10^{-6}$$

Note that the two results are slightly different by about 6%.

■ **Problem 4.17. Finding $\tau$ given the maximum observed count rate of a paralyzable detector.**

We know the maximum value of m is 50000 counts/sec, and we also know that $m = ne^{-n\tau}$. We just need to find the value of n that maximizes m. This calculation is shown below, where we differentiate the equation for m with respect to n, set it equal to zero, and solve for n.

$$\frac{\partial (n\,e^{-n\tau})}{\partial n} = 0$$

The value of n that maximizes m in terms of $\tau$ is thus:

$$n = \frac{1}{\tau}$$

Indeed, this tells us that the maximum observed count rate occurs when the true event rate is $1/\tau$. Next, we substitute this value for n and solve for $\tau$, plugging in the maximum value of m.

$$m = n\,e^{-n\tau}$$

We substitute $n = \frac{1}{\tau}$ and m = 50000 to get the dead time of the detector in seconds.

$$\tau = 7.358 \times 10^{-6}$$

■ **Supplemental Problem on Moving Sources**

■ **Moving Sources**

A cylindrical detector of area A =100 cm2 and intrinsic efficiency =100% is placed a distance d = 3 m from a road. The detector alarms if any passing vehicle causes 100 counts above background. The vehicles move at a constant speed v=3 m/s and are in view of the detector when they pass from x = -L until x = +L. Estimate the minimum activity for the alarm level if L is chosen to be a large value. [Neglect background and attenuation].

Assume $\Omega = \frac{A\cos\theta}{r^2}$ so the number of counts when the source is at x(t) is $dC(t) = \frac{S\Omega(t)\,\epsilon_i}{4\pi}\,dt$

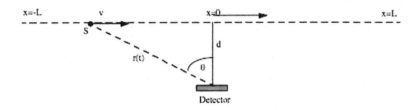

We let $r^2 = x^2 + d^2$, so

$$dC(x) = S\,A\,d\,dx \Big/ \left(4\pi \left(x^2 + d^2\right)^{(3/2)}\right) v.$$

where we substituted dt = dx/v, and used from geometry, $\cos(\theta) = d/r$. To find the total counts accumulated, we integrate dC(x) from - L to L :

$$C = \int_{-\infty}^{\infty} \frac{S\,A\,d}{4\pi v \left(d^2 + x^2\right)^{3/2}}\,dx \qquad \text{assuming } d > 0$$

$$C = \frac{A\,S}{2\,d\,\pi\,v}$$

Solving the above equation for S and plugging in known values for C, d,v and A gives us:

$$S = \frac{2\,C\,d\,\pi\,v}{A} \qquad \text{where we substitute } C = 100,\ d = 3\,m,\ v = 3\,m/s,\ \text{and } A = 0.01\,m^2$$

$$S = 565\,487\ \text{sec}^{-1}$$

With *a* spherical detector, the $\cos(\theta)$ term disappears, making the integral :

$$C = \int_{-\infty}^{\infty} \frac{S\,A}{4\pi v \left(x^2 + d^2\right)^{2/2}}\,dx \qquad \text{assuming } d > 0$$

$$C = \frac{A\,S}{4\,d\,v}$$

Solving the above equation for S and plugging in known values for C, d,v and A gives us:

$$S = \frac{4\,C\,d\,v}{A} \qquad \text{where we substitute } C = 100,\ d = 3\,m,\ v = 3\,m/s,\ \text{and } A = 0.01\,m^2$$

$$S = 360\,000\ \text{sec}^{-1}$$

This makes sense. The system can detect a smaller source because the detector is more efficient by virtue of the full area of the detector being exposed to the source at all times. For finite viewing of the source from -L to L :

$$C = \int_{-L}^{L} \frac{S\,A\,d}{4\pi v \left(d^2 + x^2\right)^{3/2}}\,dx \qquad \text{assuming } d > 0 \text{ and } L > 0$$

$$C = \frac{A\,L\,S}{2\,d\,\sqrt{d^2 + L^2}\ \pi v}$$

An interesting exercise left to the reader is to plot S as L is varied from small to large.

___

## More on the Fano Factor

The Fano factor, F, is defined as the variance of the pulse height (or energy) distribution divided by its mean. If Poisson or Gaussian statistics applied, the Fano factor should be 1. And yet we note that F is about 0.1 for semiconductor detectors. This note looks at the rationale for this large difference.

## ▪ Charge formation

The primary electron (perhaps a photoelectron) moves through the detector material and interacts with the atoms along its path. Suppose N of these interactions result in an electron-hole pair being formed. Each of these interactions will result in a different amount of energy being lost by the primary electron. (We add any non-charge-producing energy losses from other interactions to this energy loss so that the total energy lost equals the primary electron energy loss).

Suppose the distribution of the amount of energy lost in an interaction that produces a signal carrier is given by g(E). One simple example of g(E) is shown below where $E_g$ is the band gap energy.

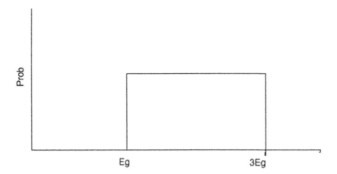

## ▪ **Combination of multiple interactions drawn from g(E)**

We can view our problem as either fixing the number of interactions, N, and calculating the probability distribution of the total energy lost $E_{tot}$ , or fixing the energy lost and calculating the probability distribution of the number of interactions. We'll choose the former for now.

If we look at all of the interactions leading to a carrier, we have a set of N random variables {$\Delta E_1$, $\Delta E_2$, ...., $\Delta E_N$). where each of these is drawn from the distribution g(E). But we seek the distribution for the total energy deposition, i.e., of the sum $E_{tot} = \Delta E_1 + \Delta E_2 + .... + \Delta E_N$.

The Central Limit Theorem immediately gives us the distribution for $E_{tot}$ assuming N is large. This fundamental theorem of statistics states that the combination of independent and identically-distributed random variables will follow a normal distribution which has a mean of N<E> and a variance of N $\sigma_{g(E)}^2$ . The reason for this is that we have N independent variables which can be combined in multiple ways to total $E_{tot}$. The interested student should consult any statistics textbook (or just see the example at http://wiki.stat.ucla.edu/socr/index.php/SOCR_EduMaterials_Activities_GeneralCentralLimitTheorem) for further elaboration.

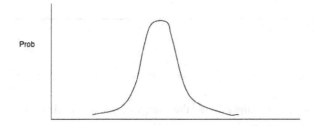

Using this result, we can immediately write down the Fano factor:

$$F = \frac{\text{var}(E_{\text{tot}})}{<E_{\text{tot}}>} = \frac{N\,\sigma_g^{\,2}}{N<g(E)>} = \frac{\sigma_g^{\,2}}{<g(E)>}$$

This tells us that the variance of the pulse height is actually due to the underlying distribution $g(E)$. If $g(E)$ happens to be Gaussian or Poisson, then F=1. For the case where the total energy is fixed, and N varies, a very similar argument can be made, and the same result for F will emerge. (cf. Jordan, D.V., et al., NIMA 585 (2008) 146-154).

- **Extra Problem**

1. Calculate the Fano factor using $g(E)$ shown in the figure above except allow $g(E)$ to be uniform over the variable range from $E_g$ to $N*E_g$. Although this is a crude model, compare your results for N=5 to the Fano factor for typical semiconductor (Knoll: Table 11.1, 2nd Ed) and gas-filled detectors (Knoll: Table 6.2, 2nd Ed.). Plot the Fano factor for N ranging from 1 to 10. Your plot shows generally how the Fano factor changes as $g(E)$ broadens.

(As an aside, Fano factors are usually not applied to scintillators, although this is a current topic of discussion. If we follow the arguments above, the primary electron produces excited atoms, each of which may produce a photon. If we could count these photons directly, then the Fano factor may have some relevance. But each photon must be collected and converted to a photoelectron. One might expect that the net result of these additional random processes is to ultimately produce a $g(E)$ distribution for the final signal carriers emerging from the PMT that is Gaussian in shape, which would imply F~1).

## Ionization Chamber problems

- **Problem 5.1. Charge created by alpha particles in He.**

The W-value (energy dissipation/ion-pair) for $\alpha$ particles in He is 42.7 eV. We simply divide the energy of one $\alpha$ particle (5.5 MeV) by the W-value to find the number of charge carriers created when the $\alpha$ particle is stopped in He:

$$N = \frac{5.5\,\text{MeV}}{\dfrac{42.7\,\text{eV}}{\text{Carrier}}}$$

The number of charge carriers created (electrons or positive ions) is thus:

$$N = 128.8\,\text{Kilo Carriers}$$

Next, we simply convert the number of carriers created to charge by multiplying by the charge of one electron.

$$Q = \frac{128.8\,\text{Kilo Carriers} \times \text{ElectronCharge}}{\text{Carriers}}$$

The charge created when one 5.5 MeV $\alpha$ particle deposits all its energy in He is thus:

$$Q = 2.06 \times 10^{-14}\,\text{Coulombs}$$

To find the corresponding saturated current when 300 $\alpha$ particles per second enter the He filled ion chamber, we simply take the number of charge carriers created by one $\alpha$ particle, and multiply that by the charge of one electron and the factor 300/second:

$$I = \frac{(128.8\,\text{Kilo Carriers} \times \text{ElectronCharge})\,300}{\text{Carriers Second}}$$

The saturated current created when 300 $\alpha$ particles per second (each with energy 5.5 MeV) enter the He filled ion chamber and deposit their full energy (in pA) is thus:

$$I = 6.19\,\text{pA}$$

Note that this is an extremely small current to be measured.

- **Problem 5.3. Finding the gamma-ray exposure necessary to reduce ion chamber voltage.**

Exposure is found by determining the charge formed (and lost from the capacitor) per volume of air at STP, and then converting that into charge/mass air, which is exposure. We define our equation for the change in charge per volume air with the voltage drop, i.e. $\dfrac{\Delta Q}{\text{Volume Air}} = \dfrac{C\Delta V}{\text{Volume Air}}$ (for Volume Air, we use the factor $\rho_{air} = 1.293 \times 10^{-3}\,g/cm^3$ to find the mass of the air, which we need for units to cancel), and convert to roentgen (R) using the factor $R = 2.58 \times 10^{-7}\,C/g$. We then take our result and plug in the factor 1 Farad-V = 1 C.

$$\frac{C\left(v_0 - v_1\right)R}{\dfrac{\left(50\,\text{cm}^3\right)\left(1.293\,\text{grams}\right)\left(2.58\,\text{Coulombs}\right)}{\left(10^3\,\text{cm}^3\right)\left(10^7\,\text{grams}\right)}}$$

We substitute $C = \dfrac{75\,\text{Farads}}{10^{12}}$, $v_0 = 25$ Volts and $v_1 = 20$ Volts to get the gamma-ray exposure in roentgen (R):

$$\text{Exposure} \quad = 0.0225\ R$$

### Problem 5.5. Finding average current over gamma-ray exposure period.

This is similar to problem 5.3. We find the charge generated corresponding to the voltage drop on the chamber, and dividing this by the exposure time, gives the average current. We define the average current as $\frac{\Delta Q}{\Delta T} = \frac{C\Delta V}{\Delta T}$ (denoted by "i" below) and give the appropriate unit conversions so that we get our result in amperes.:

$$i = \frac{c\left(v_0 - v_1\right)}{30 \times 60 \text{ seconds}}$$

We substitute $c = \frac{250 \text{ Farad}}{10^{12}}$, $v_0 = 1000$ Volts and $v_1 = 850$ Volts to get the average current over the exposure period:

$$i = 2.083 \times 10^{-11} \text{ Amperes}$$

### Problem 5.7. Compensation of free air chambers for high energy gamma-rays.

The key idea of the 1-D Free Air ion chamber is that the range of the secondary electron is less than twice the electrode separation distance. Under the worst case, a 5 MeV gamma will lead to a 5 MeV electron through a photoelectric absorption. Using the textbook Fig. 2.14, we estimate $\rho R \sim 3\ \text{gm}/\text{cm}^2$ for 5 MeV electrons. Thus, we estimate the range of these electrons using $\rho_{\text{air}} = .001293\ g/\text{cm}^3$ as in Problem 5.6. The estimated electron range, and hence roughly half the spacing needed between electrodes (in cm) is thus:

$$\text{Range} = \frac{3.0}{0.001293} = 2320 \text{ cm}$$

This implies a separation distance of $\sim 50$ meters! Not very attractive. Fortunately, a photoelectric absorption is not very likely at these energies. Nevertheless, pair production will still yield electron positron pairs with $\sim 2$ MeV of energy each. This would still require large separations of several meters between electrodes.

### Problem 5.9. Current in an ion chamber at low dose rate.

We need to find the current produced in 1 liter of STP air if the dose rate is 0.5 mR/h (35.8 pC/kg-s). First, we find the mass of 1 liter of air using its density, which is shown below.

$$m = \frac{(0.001293 \text{ kg}) \, (1000 \text{ cm}^3) \text{ liter}}{\left(10^3 \text{ cm}^3\right) \text{ liter}}$$

The mass of 1 liter of air is thus:

$$m = 0.00129 \text{ kg}$$

Then we multiply this mass of air by the dose rate in pC/kg-s to get the minimum current that must be measured.:

$$i = 0.001293 \text{ kg} \left(\frac{35.8 \text{ Coulombs}}{10^{12} \text{ (kg-sec)}}\right) = 4.63 \times 10^{-14} \text{ Amps}$$

The answer tells us that this is a very small current -- probably too small to measure reliably. We need much bigger signals (x $10^6$ bigger) to measure this level of dose rate.

### ■ Problem 5.11. Electronic equilibrium

a) According to Table 5-2, electronic equilibrium is reached for a wall thickness of 0.43 $g/cm^2$.

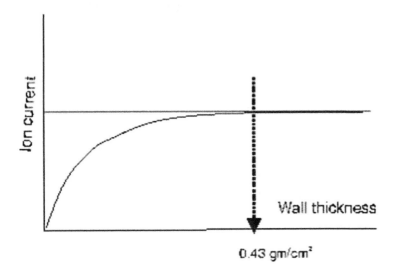

b) In a vacuum, no secondary electrons can enter the chamber, so there is inadequate compensation just using thin walls. This reduces the measured ion current compared with air surroundings.

### ■ Problem 5.13. Ion current predicted from exposure rate.

We use the ideal gas law to find the mass of air in the chamber. Using the relation that pressure is proportional to density times temperature, $p \propto \rho * T$:

$$\rho = \left(\frac{p}{p_0}\right)\left(\frac{T_0}{T}\right)\rho_0$$

we are able to find the air density at the specified conditions (the subscript "0" refers to STP conditions, and "p" and "T" refer to pressure and temperature, respectively). From this, we can find the mass of the air in the ion chamber, which we multiply by the gamma-ray exposure rate to find the saturated ion current.

First, we define the value of the exposure rate:

$$\textbf{exposure rate} = \frac{\textbf{100 pC}}{\textbf{kg-sec}}$$

And we calculate the mass of the air using the method outlined above:

$$\textbf{mass}_{\textbf{air}} = \frac{p\, t_0\, \rho_0 \times \textbf{volume}}{p_0\ t}$$

We substitute p = 3 Atm, $p_0$ = 1 Atm, $T_0$ = 273 $K$, $T$ = 373 $K$, $\rho_0 = \frac{1.293 \text{ mg}}{\text{cm}^3}$ and volume = 2500 cm³ to get the mass of air in the ion chamber:

$$\text{mass}_{\text{air}} = 7.098 \text{ g}$$

Now we calculate the saturated ion current by multiplying the exposure rate by the mass of air in the ion chamber:

$$i_{\text{saturated}} = \text{exposure rate} \times \text{mass}_{\text{air}} = 0.7098 \text{ pA}$$

### ■ Problem 5.15. Rise time slopes for electrons and ions.

We need the slope of the curves for electron + ion travel, $V_{\text{ei}}(t)$ (cf. textbook Eqn. 5.15), and for just ion travel, $V_i(t)$ (cf. textbook Eqn. 5.16). Recall that a particle's velocity is given by $\frac{\mu E}{p}$ (Eqn. 5.3), where $\mu$ is the mobility and $\frac{E}{p}$ is the electric field strength divided by the gas pressure. In the third line, we calculate the ratio of the derivative (with respect to t) of $V_{\text{ei}}(t)$ to that of $V_i(t)$, giving the appropriate relationships for $v^+$ and $v^-$.

We express Eqn. 5.15

$$V_{\text{ei}} = \frac{n_0 \, e \, (v^+ + v^-) \, t}{d \, c}$$

We express Eqn. 5.16 as

$$V_i = \frac{n_0 \, e \, (t \, v^+ + x)}{d \, c}$$

Calculate the ratio of the derivative (with respect to t) of $V_{\text{ei}}(t)$ to that of $V_i(t)$, giving the appropriate relationships for $v^+$ and $v^-$.

$$\frac{\left( \frac{\partial V_{\text{ei}}}{\partial t} \right)}{\left( \frac{\partial V_i}{\partial t} \right)}$$

where we substitute $v^+ = \mu^+ E$ and $v^- = \mu^- E$ to get the equation for the ratio of the slopes of the $V_{\text{ei}}$ and $V_i$ portions of Fig. 5.16.

$$\frac{\mu^- + \mu^+}{\mu^+}$$

Now we take our previous result and plug in estimated values for the mobilities (read pg. 135), $\mu^- = 1000 \, \mu^+$ and $\mu^+ = \frac{1.25 \text{ m}^2 \text{ Atm}}{10^4 \text{ (Volt– Second)}}$, to get the approximate ratio of the slopes of the $V_{\text{ei}}$ and $V_i$ portions of Fig. 5.16.

$$\text{Slope Ratio} = 1001$$

The rise time is proportional to the velocity, which is proportional to the carrier mobility. So the ratio of the fast (electron-dominated) rise to the slow (ion-only) rise is about the ratio of the mobilities, which is about $10^3$. Note that we have assumed that the ions aren't moving much during the first part of the pulse. Otherwise, we would need to include them, and the ratio would more precisely be given by $1 + \frac{\mu^-}{\mu^+}$, which is what we have found in our result.

## Supplemental Problem: Minimum pulse rise time

Assume for simplicity a parallel plate geometry so that the electric field is constant in the detector volume. Assume that electrons and ions move with saturated velocities $v_e$ and $v_i$. The distance between the negatively charged cathode and positively charged anode is d.

Show that the minimum rise time of the resulting pulse occurs when $\frac{x}{d} = \frac{v_e}{v_e+v_i}$ and is $t_{rise}^{min} = \frac{d}{v_{total}}$ where $v_{total} = v_e + v_i$.

Solution: Plot $t_{rise}$ (= distance to collection/velocity) as a function of interaction position x for both the electrons and ions. The plot will form an "X", with the upper "v" of the "X" shape representing the total charge collection time as a function of x. The point of intersection of the two curves is $\left(x_{rise}^{min}, t_{rise}^{min}\right)$. The reader can extend this problem to include other electric field distributions (e.g., cylindrical, spherical).

# Proportional counter problems

## ■ Problem 6.1. Variance in charge carriers without multiplication.

We expect the mean number of ion pairs formed to be equal to the energy of the radiation divided by the W-value of argon, and we expect the standard deviation in the number of ion pairs to be the square root of the Fano factor times the calculated mean number of ion pairs. Also, the relative standard deviation is simply the standard deviation divided by the mean. In the following calculation Fano = 0.17. We express this below.

$$\text{mean number of ion pairs} = \frac{10^6 \text{ eV}}{26.2 \text{ eV}} = 38\,170$$

$$\text{Expected standard deviation } (\sigma) = \sqrt{\text{Fano} \times \text{mean}} = 80.6$$

$$\text{Relative std deviation} = \frac{\sigma}{\text{mean}} = 0.00211$$

Now we look at what happens if the Fano factor equals 1.

$$\text{mean number of ion pairs} = \frac{10^6 \text{ eV}}{26.2 \text{ eV}} = 38\,170$$

$$\text{Expected standard deviation } (\sigma) = \sqrt{\text{Fano} \times \text{mean}} = 195$$

$$\text{Relative std deviation} = \frac{\sigma}{\text{mean}} = 0.00512$$

Observe that the Fano factor makes a **large** difference in reconciling experiment to theory in this case, primarily because the non-Poisson recombination along the tracks reduces the fluctuation (standard deviation) in the number of ion pairs formed. (For the inquiring reader, earlier in this solutions manual is a simple explanation of the basis for the Fano factor).

## ■ Diethorn Model.

The Diethorn model plays a role in the following problems. We define it here as "multiplication", and use it throughout the following problems.

$$\text{multiplication}(V, b, a, \Delta V, K, p) = e^{\dfrac{V \ln(2) \ln\left(\dfrac{V}{K\, p\, a \ln\left(\frac{b}{a}\right)}\right)}{\ln\left(\frac{b}{a}\right) \Delta V}}$$

## ■ Problem 6.3. Effect of anode/cathode radius on multiplication.

a) We solve this part by simply giving the known values of our variables in our function "multiplication," setting "multiplication" equal to 1000, and solving for "V". Our known values in this problem are: b=1, a=.003, $\Delta V$=23.6, K=4.8×10$^4$, and p=1.

The operating voltage (in Volts) required to achieve a gas multiplication factor of 1000 is thus:

$$V = 1790 \text{ volts}$$

b) Here, we look for the effect of doubling the anode radius at this operating voltage on the multiplication factor. We calculate the factor that the multiplication changes by taking the value of "multiplication" at double the anode radius (.006 cm) and divide it by the value of "multiplication" at the previous anode radius (.003 cm). The factor the multiplication changes by (or is "increased" by) is thus:

$$\frac{\text{multiplication}\left(V, 1, .006, 23.6, 4.8\times10^4, 1\right)}{\text{multiplication}\left(V, 1, .003, 23.6, 4.8\times10^4, 1\right)} = 0.00752$$

We just take the inverse of the previous result to get the factor the multiplication is decreased by when changing the anode radius from .003 to .006 cm.

$$0.007522^{-1} = 133$$

Note that increasing the anode radius by just 30 $\mu$m decreases the multiplication by more than 100!

c) Here, we do the same as in part b, but this time we are looking at the effect of changing the cathode radius from 1 to 2 cm (we put the anode radius back to the original .003 cm). The factor the multiplication changes by is thus:

$$\frac{\text{multiplication}\left(V, 2, .003, 23.6, 4.8\times10^4, 1\right)}{\text{multiplication}\left(V, 1, .003, 23.6, 4.8\times10^4, 1\right)} = 0.192$$

Again, we take the inverse of the previous result to get the factor the multiplication is decreased by when changing the cathode radius from 1 to 2 cm.

$$0.192226^{-1} = 5.202$$

The effect is only a factor of 5 as opposed to more than 100, showing how the anode radius is quite critical in determining the multiplication.

■ **Problem 6.5. Testing of proportionality of tube.**

Using different, but known, energy depositions, measure the resulting pulse amplitudes. If the ratio of pulse amplitudes is consistent with the ratio of energy depositions, the tube is operating in proportional mode.

■ **Problem 6.7. Gas multiplication needed for given pulse height desired.**

This is the same type of problem as Problem 6.4, except here we need to find the gas multiplication which yields a pulse amplitude of 50 mV for 50 keV input. This is expressed below, where we solve the equation V= Q/C for the multiplication, "M."

$$10 \text{ Milli Volts} = \frac{M\left(5\times10^4 \text{ eV}\right)1.602 \text{ x } 10^{-19} \text{ Coulombs}}{(26 \text{ eV})(200 \text{ pF})}$$

Solving for the multiplication, M, yields:

$$M = 6490$$

■ **Problem 6.9. Field tube function.**

The function of the field tube is to reduce the electric field near the ends of the tube where distortions in the electric field would occur (particularly near corners).

## Problem 6.11. Gas scintillation detector voltage.

In gas scintillators, our signal carriers are the photons generated from the electrons moving in the high field region of the tube (around the anode). If we operated the detector in the proportional regime, we would increase the number of electrons (and hence photons), but would suffer from the loss of energy resolution associated with gas multiplication. In order to avoid this energy resolution loss, the tube is normally operated below the threshold for multiplication.

- ## Supplemental Problem: Electric Field Distributions for Cylinders and Spheres

For the Townsend coefficient $\alpha$ to be positive, the electric field must exceed $\sim 10^6$ V/m. Above this value, the electron carriers will undergo multiplication. Given that a fixed voltage $V_{appl}$ is to be placed on a central anode of radius $r_i$, with a cathode at radius $r_o$, determine the fraction of the detector volume that will undergo multiplication for spherical and cylindrical geometries. Comment on which geometry will require the least voltage to achieve multiplication. Use reasonable values for any parameters needed, such as $V_{appl} = 10^3$ V, $r_i = 10^{-5}\,m$, $r_0 = 10^{-2}\,m$.

[The motivated reader will solve the homogeneous Gauss's Law for the electric field distribution using the appropriate coordinate system. The result will show that the radius at which the electric field is at the critical value, $r_{crit}$, is similar for both geometries, but the fraction of volume that is active is much smaller for the spherical geometry].

# GM counter problems

## ■ Problem 7.1. GM quench gas ionization potential

The purpose of the quench gas in a GM tube is solely to absorb the charge from a major fill gas ion to prevent an electron from being liberated when that ion hits the cathode wall. For the quench gas to give up an electron to the positive ion, it must have a lower ionization potential than the ion.

## ■ Problem 7.3. GM Pulse height increases with V.

It is the space charge buildup around the anode wire that limits the size of the GM pulse. This limit is achieved later (higher multiplication) with increased electric fields in the detector. Since increased applied voltage implies increased electric field, the pulse height will be larger with higher applied voltages.

## ■ Problem 7.5. Proportional versus Geiger tubes.

a) <u>Proportional</u>: The pulse height varies as the avalanche amplitude which, in turn, depends on voltage in an approximately exponential manner.

 <u>Geiger</u>: The pulse amplitude corresponds to the number of ion pairs at the point at which the accumulated positive space charge is sufficient to reduce the electric field below its critical value. This number will increase in approximate proportion to the original electric field or linearly with the applied voltage.

b) <u>Proportional</u>: The quench gas must absorb UV photons.

 <u>Geiger</u>: The quench gas must pick up positive charges from the original positive ions through charge transfer collisions.

c) <u>Proportional</u>: Because heavy charged particles tend to deposit all of the energy, and electrons only part of theirs, the two radiations can be separated by their different pulse heights.

 <u>Geiger</u>: No differentiation can be achieved, because pulse height is independent of particle type and energy.

d) <u>Proportional</u>: The maximum counting rate is often set by pulse pile-up. The minimum pulse shaping time (that will minimize pile-up) is limited by the finite rise time of the pulses.

 <u>Geiger</u>: The maximum counting rate is limited by the long dead time of the tube itself.

e) <u>Proportional</u>: Gamma rays produce very small amplitude pulses and are often below the discrimination level.

 <u>Geiger</u>: Counting efficiency is a few percent due primarily to the liberation of secondary electrons from the detector walls.

## ■ Problem 7.7 GM Tubes paralyzable or non-paralyzable?

The GM tube has historically been treated as a paralyzable detector. After an avalance, the positive ions produced around the anode wire must be cleared to the cathode to re-establish the full electric field in the tube. As this positive sheath moves to the cathode and the electric field begins to be restored, another interaction in tube can initiate a smaller avalanche and require the clearing process to begin again. If the counting system requires a full discharge to register a count, then the deadtime continues to be extended with each additional interaction in the tube. Extendable deadtimes are the hallmark of the paralyzable model.

## Problem 7.9. Dead time losses in GM tubes

We are given a GM tube with a resolving time of 350 $\mu$s. We want to find the true rate for which the measured count rate is 1/2 the true rate, using both paralyzable and non-paralyzable cases.

Recall the resolving time is the time required for the GM tube to rest after a measured event before another pulse can be recognized (above some finite threshold). The dead time is the time during which no pulse of any finite amplitude can be generated (i.e. shorter than the resolving time), and the recovery time is the time from pulse onset until another full pulse can be generated (i.e. much longer than the resolving time).

First, we look at the non-paralyzable case. To find the true rate, n, we solve Eqn. 4.23 similtaneously with m = $\frac{n}{2}$ (given) for n by eliminating m. Our two expressions are thus:

$$n - m = n\, m\, \tau \qquad\qquad m = \frac{n}{2}$$

We get the trivial solution n = 0 and the solution we are interested in, n = $\frac{1}{\tau}$.

We take the second solution from above and plug in our known value $\tau = 0.000350$ to get the true rate for the non-paralyzable case in $\sec^{-1}$ of:

$$n = 2857 \ \sec^{-1}$$

Now, we look at the paralyzable case, doing the same thing as before, except using Eqn. 4.27. The two expressions are now:

$$m = n\, e^{-n\tau} \qquad\qquad m = \frac{n}{2}$$

Again, we get the trivial solution n = 0 and the solution we are interested in.

$$n = 0 \qquad \text{and} \qquad n = \frac{\ln(2)}{\tau}$$

We take the second solution from above and plug in our known value $\tau = 0.000350$ to get the true rate for the paralyzable case in $\sec^{-1}$:

$$n = 1980 \ \sec^{-1}$$

Note that the limit is reached sooner for the paralyzable case, but not by that much. Also note that these detectors will be useful only up to a couple of thousand of counts/sec.

# Scintillation detector problems.

## ▪ Problem 8.1. Scintillation efficiency.

We calculate the efficiency by multiplying the number of photons created by the energy per photon, and divide this by the total energy deposited.

Using Photon Energy = Planck's Constant x Speed of Light / Wavelength:

$$\epsilon = \frac{20\,300 \times \text{Planck' } s \text{ Constant} \times c}{(447 \text{ nm})\left(10^6 \text{ eV}\right)}$$

The scintillation efficiency expressed as a percent is thus:

$$\epsilon \quad = 5.63\,\%$$

This is actually a fairly inefficient process in converting incident energy into light energy. More importantly is the number of carriers created -- ~ 20,000 for 1 MeV. Next, we approximate the number of carriers created per eV of deposited energy:

$$\frac{20\,000 \text{ carriers}}{10^6 \text{ eV}} = \frac{0.02 \text{ carriers}}{\text{eV}}$$

We'd like to relate this to the W-value in gas-filled detectors, so we take the inverse of the previous result to get the approximate amount of energy in eV required to create one carrier.

$$= \frac{50 \text{ eV}}{\text{carrier}}$$

This is even worse than for a gas-filled detector (~ 30 eV/carrier). Fortunately, sodium iodide, the workhorse of scintillators, is much better than anthracene.

## ▪ Problem 8.3. Maximum brightness of scintillator.

The brightness is the rate of photon emission. We assume that the excited states are formed instantaneously, followed by scintillation governed by a single time constant $\tau$, so that $\frac{dn}{dt} = \frac{-1}{\tau} n = I(t) = \frac{-n_0}{\tau} e^{\frac{-t}{\tau}}$. Since I(t=0) is the largest value, the ratio of maximum brightness between NaI(Tl) and anthracene will be given by $\left(\frac{n_{0,\,NaI(Tl)}}{n_{0,\,anthra}}\right) * \left(\frac{\tau_{anthra}}{\tau_{NaI(Tl)}}\right)$. This is shown in the following expression using values from Tables 8.1 and 8.3:

$$\frac{n_{0,\,NaI(Tl)}}{n_{0,\,anthra}} \quad \frac{\tau_{anthra}}{\tau_{NaI(Tl)}}$$

We substitute $n_{0,NaI(Tl)} = 230$, $n_{0,anthra} = 100$, $\tau_{anthra} = 0.030$ and $\tau_{NaI(Tl)} = 0.23$ to get the ratio of maximum brightness between NaI(Tl) and anthracene:

$$\text{maximum brightness NaI(Tl) /Anthracene} \quad = 0.3$$

Note that while the light output is 2.3 times larger for NaI, the brightness also depends on how fast the light comes out, which is much faster for anthracene. Thus, anthracene is brighter (maximum light emission rate) by a factor of ~3.

## ■ Problem 8.5. The role of activators.

The activator in an inorganic scintillator is the atom impurity that actually produces the detectable scintillation light. Its energy structure lies in the forbidden energy gap of the host material. When an electron is excited to the conduction band in the host material, it migrates to an activator site and non-radiatively loses energy so that it occupies the lowest level of the activator's excited state. The subsequent de-excitation is at a wavelength that is too long to cause ionization in the bulk material, and this scintillation photon flows freely to a photodetector for detection.

In an inorganic material, the excitation and de-excitation occur with the molecule itself.

## ■ Problem 8.7. Scintillation Mechanisms.

This problem is similar to problem 8-5. The key is to recognize that in an inorganic scintillator, a crystal lattice is required to establish the energy levels and the activator (a deliberate impurity) is chosen to have energy levels that lie within the forbidden energy gap of the host material. Without the crystal structure present, there would be no scintillation possible.

In contrast, inorganic scintillators function using the energy levels of the molecules themselves. Thus, a single molecule can function as a scintillator, and does not require a crystalline structure.

## ■ Problem 8.9. Photons generated with 1 MeV deposition in NaI.

Using data from p. 237, we calculate that 40,000 photons are created per MeV deposited in NaI(Tl) emitted isotropically from the following relation: $\frac{(\text{Scint.eff .for 1 MeV } \beta)(1\text{ MeV})}{(hc/\lambda)} = \frac{12\,\%*1\text{ MeV}}{3\text{ eV}}$ which is close to the value of 38,000 in Table 8.3. The solid angle subtended by the pupil is $\Omega \sim \frac{A}{d^2}$ (look at Eqn. 4.22; we assume d>>a), and probability of detection (or the probability that an emitted photon hits the pupil in this case) is $\frac{\Omega}{4\pi}$. Therefore, the expected number of photons that will hit the pupil per event are (43,300 photons)*($\frac{\Omega}{4\pi}$). This is expressed below:

$$\text{Number of photons we expect to hit the pupil per event} = \frac{40\,000\,\pi\,(1.5\text{ mm})^2}{4\,\pi\,(10\text{ cm})^2} = 2.25$$

This is not sufficient for detection by the human eye (we need at least 10 photons to hit the pupil per event).

### ■ Supplemental Problem: Grid-less GPSC using single carrier approach

The gas proportional scintillation counter (GPSC) is commonly used in the detection of X-rays. In a conventional GPSC, the x-rays interact in a drift/absorption region having an electric field below the scintillation and ionization thresholds.

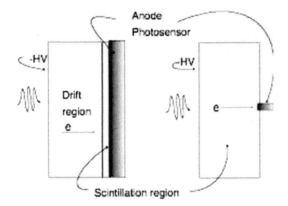

They drift past a grid into a thin higher electric field region where they then cause scintillation, but not ionization, which is detected by the anode photosensor. In the new gridless detectors, the entire volume is operated at a voltage where electrons cause scintillation during their entire travel distance, but the photosensor now is chosen to be small enough to subtend a small solid angle until the electron-induced scintillations occur close to the anode.

For both detectors, derive expressions for and sketch the shape of the observed anode signal as a function of the x-ray interaction position $z_0$ and time t. For simplicity you may assume: the electrons move directly to the anode with constant velocity $v_0$, a constant light generation per electron path length k in the scintillation region, and a cylindrical photosensor of area A. For simplicity, you may assume $\Omega = \frac{A}{z(t)^2}$ for z > A.

The electrons move with velocity v after being formed at position z away from the photosensor. They isotropically produce scintillation light at a rate of k photons/path-length, which is detected by a cylindrical photosensor of area A and radius a. The sensed light per path length of electron travel is given by:

$S(z) = \frac{k\Omega}{4\pi}$

Assume that the electron is formed on the axis of the photosensor and moves directly to it, then

$$\frac{\Omega}{4\pi} = \frac{1}{2}\left(1 - \frac{d}{\sqrt{d^2+a^2}}\right) = \frac{1}{2}\left(1 - \frac{z-vt}{\sqrt{(z-vt)^2+a^2}}\right)$$

Since S(z)dz=S(t)dt, S(t)= v S(z). Let's assume some unit dimensions (k = v = z = a = 1), and write

$$S(t) = \frac{1}{2}(k\,v)\left(1 - \frac{z-t\,v}{\sqrt{a^2+(z-t\,v)^2}}\right)$$

First, ignore the grid and assume that the electron scintillates all the way to the photosensor (i.e., choose z~a). A plot of S(t) yields:

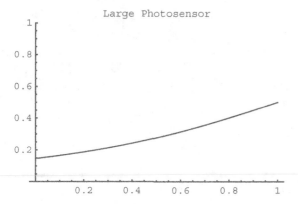

We have a pulse that increases from $z_0$ to the end. The pulse will be dependent upon the electron's formation in the chamber.

Next, assume that the electron only radiates only after passing through the grid, so z<<a. A plot of the output for distance greater than 90% is:

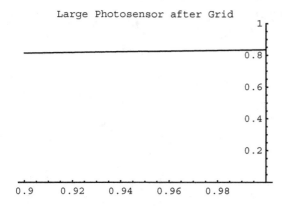

This is like a Frisch grid ionization chamber. Any electron first travels to the grid, then produces its signal in the small region between grid and photosensor. Therefore, each electron, no matter where it is formed, creates the same size pulse. This is what is needed for spectroscopy. The disadvantage of a grid is that it is typically noisy, and there may be some electrons that are initially formed in between the grid and photosensor.

Finally, assume a small photosensor without a grid. We assume a=0.1, and the full travel distance. The S(t) output looks like:

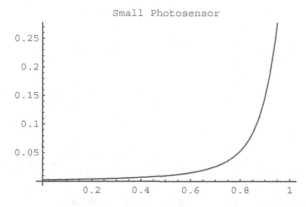

This is analogous to the "single carrier approach" that is popular today. The electron produces most of its signal only when it is very close to being collected, therefore the signal is largely independent of the formation position of the electron. Combining our results on one plot:

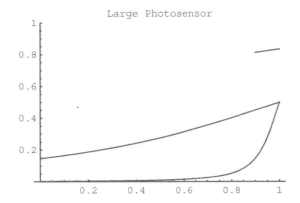

□ Further exercises:

1. For the small photosensor, let a=.01 and see how your results change.

2. From the results, which of these 3 approaches will have better:

a. energy resolution  (Consider the size of the signal and whether the signal is dependent on the position of interaction).

b. timing resolution (shorter and larger pulses have better timing)

c. spatial resolution (in the z-direction)

## PM Tube and PD Problems

■ **Problem 9.1. Cutoff wavelength due to photocathode work function.**

There will be no photoelectric effect if the incoming photon energy is less than the work function of the material, i.e. the energy of the photon must be at least equal to the work function (in this case, 1.5 eV). To solve this problem, we set the energy of the incoming photon equal to the work function and solve our equation for the photon wavelength (denoted "$\lambda$").

$$\frac{(\text{Planck's Constant}) \times c}{\lambda} = \text{work function}$$

We solve the above equation for the photon wavelength ($\lambda$) and substitute work function =1.5 eV to get the maximum photon wavelength (long-wavelength limit) in angstroms.

$$\lambda = 8270 \text{ angstroms}$$

■ **Problem 9.3. Multistage gain.**

We are given that $\delta^N = 10^6 = \delta^6$ (because we know that N = 6), which means that $\delta = 10$. From Fig. 9.3, we find that the primary electron energy must be ~ 200 eV. Therefore, the voltage between dynodes (or the voltage per stage) must be 200 V, which means that we would need 1200 V across the entire tube.

■ **Problem 9.5. Gain change with voltage fluctuation.**

A 10-stage tube with each stage having a multiplication $\delta = V_S^{0.6} = \left(\frac{V}{N}\right)^{0.6}$ operates with V= 1000 Volts ($V_S$ is the interdynode voltage). Since N=10, we have $\delta = \left(\frac{V}{10}\right)^{0.6}$. Therefore, the total gain of the tube is G= $\delta^N = \left[\left(\frac{V}{10}\right)^{0.6}\right]^N = \left[\left(\frac{V}{10}\right)^{0.6}\right]^{10} = 10^{-6} V^6$. The gain fluctuation (expressed as "$\Delta$G") depends on voltage fluctuation (expressed as "$\Delta$V") by taking the differential of the total gain with respect to the voltage (i.e. we assume that $\frac{\Delta G}{\Delta V} = \frac{dG}{dV}$ for the small change in gain we are considering). This is shown below, where we have solved for $\Delta$G.

$$\Delta G = \frac{\partial\left(\frac{V^6}{10^6}\right)}{\partial V} \Delta V = 6 \times 10^{-6} V^5 \Delta V$$

We are interested in finding $\Delta$G/G for V=1000 volts. (Note that $\Delta G = 6 \times 10^9 \Delta V$ --- so tiny changes in voltage yield large changes in gain).

Looking at the relative change in gain for V=1000 V:

$$\frac{\Delta G}{G} = \frac{\Delta G}{\left(\frac{V^6}{10^6}\right)} = \frac{6 \times 10^{-6} V^5 \Delta V}{\left(\frac{V^6}{10^6}\right)} = 6 \times 10^{-3} \Delta V$$

We want to solve this equation for the $\Delta$V which yields a relative uncertainty in the gain of 1% (i.e. $\Delta$G/G = 0.01). The voltage fluctuation (in volts) that can be tolerated if the gain is not to change by more that 1% is thus:

$$6 \times 10^{-3} \Delta V = 0.01$$

or solving for $\Delta V$

$\Delta V = 1.67$ volts

This implies that the voltage must be held constant to 1 part in $10^3$ to keep the gain constant to within 1%.

### ■ Problem 9.7. Voltage response of a finite step current into parallel RC circuit.

We must first solve the differential circuit equation (Eqn. 9.12) for v(t). This is expressed below, where we make "t" the independent variable and we give the initial value v(0) = 0 (note that we wrote "$I_0$" instead of "I(t)" because I is a constant until it abruptly changes to 0 at t=T). (In this problem $v_1$ = v(t) from $0 < t < T$, and $v_2$ = v(t) from $T < t$)

$$\frac{v_1(t)}{\tau} + v_1'(t) = \frac{I_0}{c} \qquad \text{where } v_1(0) = 0$$

The solution for $v(t)$ is thus:

$$v_1(t) = \frac{I_0\,\tau}{c} - \frac{I_0\,\tau}{c\,e^{t/\tau}}$$

We factor the right hand side of v(t) so we have one term instead of two.

$$v_1(t) = \frac{(-1 + e^{t/\tau})\,I_0\,\tau}{c\,e^{t/\tau}}$$

Now, we plot v(t) for $0 < t < T$ (we set T = 5 and $\tau = RC = 1$).

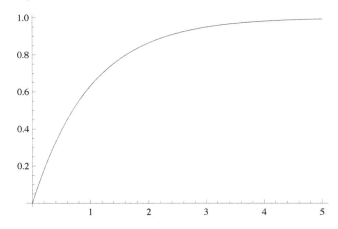

After the current stops flowing at T=5, the differential equation has i(t) =0, and the initial condition that the voltage matches at t=5 (i.e. $v_1(t = 5) = v_2$(t=5) is our initial voltage value for t = T). Below, we solve the differential equation given these conditions.

$$\frac{v_2(t)}{\tau} + v_2'(t) = 0 \qquad \text{where } v_2(5) = v_1(5)$$

The solution for $v(t)$ for t >T is thus:

$$v_2(t) = e^{-\frac{t}{\tau} + \frac{5}{\tau} - 5}\left(-1 + e^5\right)$$

The solution for $v_2$(t) in terms of t with $\tau = 1$ (t > T).

$$v_2\,(t) = \frac{-1 + e^5}{e^t}$$

Now we plot v(t) for t > T (T = 5 and $\tau$ = RC = 1).

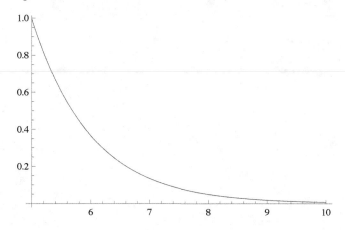

Next, we put the two solutions together on one plot to give the time profile of v(t) with $\tau$ = RC = 1 and T = 5 (and C = 1, I = 1 for 0 < t < T, and I = 0 for t > T).

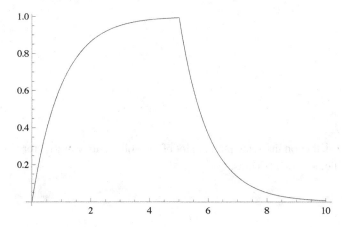

Now, we switch the values of $\tau$ = RC and T (i.e. RC = 5, T=1). We plot $v_1$(t) for 0 < t < T, where T = 1.

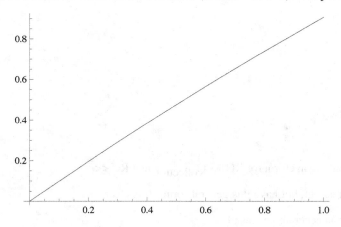

As before, we solve the differential equation for v(t) for t > T, giving the initial value of v(T) = v(1) from the previous plot. In this part of the problem T = 1, $I_0$ = 1and c = 1.

$$\frac{v_2(t)}{\tau} + v_2{}'(t) = 0 \qquad \text{where } v_2(1) = v_1(1)$$

The solution for v(t) for t >T is thus:

$$v_2(t) = 5\left(-1 + \sqrt[5]{e}\right) e^{-\frac{t}{\tau} - \frac{1}{5} + \frac{1}{\tau}} = 5\left(1 - e^{-0.2}\right) e^{-(t-1)/\tau}$$

The solution for v(t) for t > T in terms of t with $\tau$ = RC = 5.

$$v_2(t) = \frac{5\left(-1 + \sqrt[5]{e}\right)}{e^{t/5}}$$

We now plot v(t) for t > T ($\tau$ = RC = 5 and T = 1).

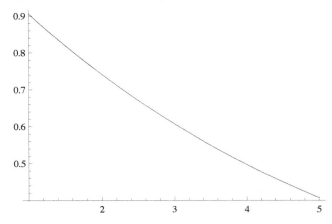

Again, we put the two sections of the voltage-time profile together on the same plot. A plot of the voltage-time profile for $\tau$ = RC = 5 and T = 1 (this is actually what part (a) was asking for, i.e. RC >> T) is thus:

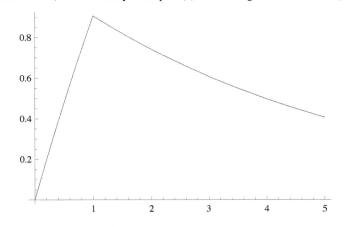

What we see here is a direct demonstration of the output pulse waveform for RC >> $t_{collection}$ and RC << $t_{collection}$. In practice, the actual wave forms are smoothed out, but have this general form. It is clear that we usually would want RC >> $t_{collection}$ so that we can accurately measure $V_{max}$.

### Problem 9.9. Microchannel plate primary advantage

Microchannel plates (MCPs) are very popular readouts for a number of reasons. They are compact and less sensitive to magnetic fields than traditional PMTs, but their primary advantage lies in their timing resolution. Because they are quite thin, the electron transit time is much shorter than in a traditional PMT and therefore the spread in arrival times at the anode, the factor which governs timing resolution, is much narrower.

### ■ Problem 9.11. Scintillator-photodiode current calculation.

To solve this, we first calculate the number of photons generated per time (alpha particle energy multiplied by a scintillation efficiency of 3%, and divided by average photon energy, which is then multiplied by the alpha particle flux, or "rate"). We then multiply that by the fraction of photons collected (light collection efficiency, or "$\epsilon_L$") and the fraction of photons generating photoelectrons (quantum efficiency, or "QE"). Finally, we take this result and multiply it by the charge of a single photoelectron to obtain the photodiode current. This is shown below.

$$\text{signal} = \frac{(3\%)(5\,\text{MeV})}{(\text{Planck's Constant})\left(\frac{c}{\lambda}\right)} \times \text{rate} \times \epsilon_L \times \text{QE} \times (\text{Electron Charge})$$

We substitute $\epsilon_L = 80\%$, $\lambda = 420\,\text{nm}$, QE = 75 % and rate = $10^6\,\text{sec}^{-1}$ to get the expected signal (or photodiode current) when the photodiode is operated in current mode.

$$\text{signal} = 4.88 \times 10^{-9}\,\text{amperes} = 4.88\,\text{nA}$$

Note that we can expect about a nanoamp per MeV deposited in the detector-PD combination. This is a very small current to measure, but can be done.

## Spectroscopy with Scintillator problems

- **Compton scattering relations.**

The energy relations for Compton scattering play a role in some of the following problems. We have defined them here for future use. The scattered photon energy $h\nu'$ is denoted as "$E_p$," the energy of the Compton electron is denoted as "$E_e$," and the incident photon energy $h\nu$ is denoted as "$E_0$." Using these definitions, our key relationships for Compton scattering are given by:

$$E_p(E_0, \theta) = \frac{E_0}{\frac{E_0(1-\cos(\theta))}{m_e c^2} + 1}$$

$$E_e(E_0, \theta) = E_0 - E_p(E_0, \theta)$$

- **Problem 10.1. Find $E_0$ given $h\nu'$ and $\theta$ in Compton scattering.**

We are given the energy of the scattered photon $h\nu'$ (or "$E_p$" from the relationships above), so all we need to do is solve for $E_0$ using our known values.

$$E_p = \frac{E_0}{\frac{E_0(1-\cos(\theta))}{m_e c^2} + 1}$$

We substitute $\theta=90°$ and $E_p=0.5$ MeV into the above equation and solve for $E_0$ to get the value of $E_0(= h\nu)$ in eV.

$$E_0 = 2.32 \times 10^7 \text{ eV}$$

- **Problem 10.3. Maximum energy deposited by two Compton scatterings.**

The largest deposited energy occurs for a 180 degree scattering. Here, we calculate the final photon energy for two 180 degree scatterings.

$$E_p\big(E_p(1\text{ MeV, } 180°), 180°\big) = 0.113 \text{ MeV}$$

The deposited energy is given by the initial photon energy minus the final photon energy. The maximum deposited energy by a 1 MeV gamma ray from two successive Compton scatterings is thus:

$$E_{dep} = 1\text{ MeV} - 0.113278\text{ MeV} = 0.887\text{ MeV}$$

Note that the photon has lost nearly **90%** of its energy with just these two scatterings.

- **Problem 10.5. Predicted peak-to-total ratio based on attenuation coefficients.**

The peak-to-total ratio will be larger than the ratio of attenuation coefficients because of histories that may begin with Compton scattering or pair production, but for which the full energy is eventually absorbed. For small detectors, it will be closer to the attenuation coefficient ratios, whereas it will be larger for larger detectors.

- **Problem 10.7. Intrinsic total efficiency of NaI(Tl) at 0.5 MeV.**

Since any interaction will give some signal, the total efficiency counts every gamma-ray which has had some interaction in the detector. Since we know that $\frac{I}{I_0}$ is the percentage of photons that don't interact in an absorber (i.e. the detector), we know that 1-$\frac{I}{I_0}$ is the percentage that do interact in the detector, which is the same as the intrinsic total efficiency. We are given the mass attenuation coefficient $\left(\frac{\mu}{\rho}\right)$ and the thickness of the detector, and the density of NaI(Tl) is found from the specific gravity in Table 8.3 to be $3.67 \ g/cm^3$, so the intrinsic total efficiency can be calculated as:

$$\text{efficiency}_{total} = 1 - e^{-\frac{\mu \rho t}{\rho}}$$

We substitute $\mu = 0.955 \ \rho$, $\rho = 3.67$, and $t = 0.5$ to get the intrinsic total efficiency of a 0.5 cm thick NaI(Tl) slab detector at a gamma-ray energy of 0.5 MeV (expressed as a fraction, not a percent).

$$\text{efficiency}_{total} = 0.827$$

Since the photofraction is the fraction of all counts in the spectrum which lie under the peak, the intrinsic peak efficiency is the intrinsic total efficiency multiplied by the photofraction. The intrinsic peak efficiency of the same detector under the same conditions (expressed as a fraction, not a percent) is thus:

$$\text{efficiency}_{peak} = 0.826647 \times 0.4 = 0.331$$

This is rather informative. It says that for 500 keV, nearly 80% of the incident gammas will have some interaction in the 0.5 cm thick detector, and nearly half of these will deposit all of their energy in the scintillator. As a result, nearly 1/3 of incident gammas at 500 keV will deposit all their energy in the scintillator. This is a high efficiency detector.

- **Problem 10.9. Factors relating to peak efficiency and energy resolution.**

Factors which influence intrinsic peak efficiency are those which affect the capture of the the full energy of the radiation in the detector: (a) density of detector, (c) atomic number.

Factors which influence energy resolution are those which affect the number of information carriers: (b) kinetic energy required to create an information carrier (i.e. photon), (e) gain of PM tube (multiplication statistics degrade energy resolution slightly, see text pg. 344), (f) quantum efficiency of photocathode, (h) light collection efficiency.

The amplifier gain has no effect on either. It may introduce a small amount of noise.

The source-detector geometry has no first-order effect on the *intrinsic* peak efficiency.

- **Problem 10.11. True and chance peak count coincidences.**

This problem looks at the area under the sum peak compared to a single peak for a source emitting two uncorrelated gamma-rays. We first recall our expression for solid angle from Chapter 4 below:

$$\Omega[d, a] = 2\pi \left( 1 - \frac{d}{\sqrt{a^2 + d^2}} \right)$$

(a). The ratio we want is the number of true coincident full-energy (i.e. sum) counts divided by the number of counts under the $\gamma_1$ peak. Note that the number of counts under the $\gamma_1$ peak is reduced by the fact that some of the $\gamma_1$'s are lost due to the coincident summing. Hence, the ratio we want is:

$$\frac{N_{12}}{N_1|_{\text{with summation}}} = \frac{N_{12}}{N_1 - N_{12}} = \frac{S\epsilon_1 \epsilon_2 y_1 y_2 \Omega_f^2}{S\epsilon_1 y_1 \Omega_f(1 - \epsilon_2 y_2 \Omega_f)} \text{(from Eqns. 10.11 and 10.12)} = \frac{\epsilon_2 y_2 \Omega_f}{(1 - \epsilon_2 y_2 \Omega_f)},$$

where $\epsilon_2$ is the intrinsic peak efficiency of $\gamma_2$, $y_2$ is the yield of $\gamma_2$ ($y_1$ and $y_2$ are both 100% in this case, so we need not worry about them in computations), and $\Omega_f$ is the fractional solid angle (i.e. $\frac{\Omega}{4\pi}$). This calculation is shown below, where $\epsilon_{2,\text{abs}} = \epsilon_2 y_2 \Omega_f$ (absolute peak efficiency).

$$\text{ratio} = \frac{\epsilon_{2,\text{abs}}}{1 - \epsilon_{2,\text{abs}}}$$

We substitute $\epsilon_{2,\text{abs}} = \frac{\Omega(10,5)\epsilon_2}{4\pi}$ and $\epsilon_2 = 0.3$ to get the ratio of counts under the sum peak ($N_{12}$) to the counts under the $\gamma_1$ peak ($N_1|_{\text{with summation}}$, or $N_1 - N_{12}$).

**ratio = 0.0161**

(b). We first want to find the sum peak count rate given S = 100 kBq. To do this, we simply use Eqn. 10.11 (noting, again, that the yields, $y_1$ and $y_2$, both equal 100%). This calculation is shown below:

$$\text{rate} = S * \epsilon_{2,\text{abs}} * \epsilon_{1,\text{abs}}$$

We substitute $\epsilon_{2,\text{abs}} = \frac{\Omega(10,5)\epsilon_2}{4\pi}$, $\epsilon_2 = 0.3$, $\epsilon_{1,\text{abs}} = \frac{\Omega(10,5)\epsilon_1}{4\pi}$, $\epsilon_1 = 0.5$ and $S = 100 \times 10^3$ to get the rate at which events are recorded in the sum peak in $\sec^{-1}$:

**rate = 41.8 sec$^{-1}$**

*What is the additional <u>chance</u> coincidence rate in the sum peak?*

This is tricky. To get a chance count in the sum peak, we must get a full-energy peak detection from one gamma-ray and NO true coincidence from the other gamma-ray, followed by another decay which yields only the full-energy peak of the other gamma-ray (again with no true coincidence allowed). These conditions are required because we want the chance coincidence count to lie in the sum peak. Thus, the rate we want is:

$2(r_1 - r_{12})(r_2 - r_{12})\tau = 2(S\epsilon_{1,\text{abs}})(1 - \epsilon_{2,\text{abs}})(S\epsilon_{2,\text{abs}})(1 - \epsilon_{1,\text{abs}})\tau$  where $\tau$ is the detector resolving time. This calculation is shown below.

$$\text{additional rate} = 2(S\epsilon_{1,\text{abs}})(1 - \epsilon_{2,\text{abs}})(S\epsilon_{2,\text{abs}})(1 - \epsilon_{1,\text{abs}})\tau$$

We substitute $\epsilon_{2,\text{abs}} = \frac{\Omega(10,5)\epsilon_2}{4\pi}$, $\epsilon_2 = 0.3$, $\epsilon_{1,\text{abs}} = \frac{\Omega(10,5)\epsilon_1}{4\pi}$, $\epsilon_1 = 0.5$, $S = 100 \times 10^3$ and $\tau = 3 \times 10^{-6}$ to get the additional rate expected from chance coincidences (in $\sec^{-1}$) for a detector with resolving time of $3\mu s$:

**additional rate = 24.0 sec$^{-1}$**

Another way to get counts under the sum peak is to have two true coincidences which sum to less than the sum peak, and then a chance coincidence which happens to sum the total deposition up into the peak. This is highly unlikely.

### Problem 10.13. Source of 511 keV gamma-rays.

The source might produce (1) positrons which produce annihilation photons, or (2) gamma-rays of sufficiently high energy to undergo pair production followed by annihilation photons in surrounding materials.

### ■ Problem 10.15 Low versus high Z scintillators

For spectroscopy, one wants the full energy of the incident radiation deposited within the detector material. In order to promote interactions with gamma rays, a high Z material is needed since all of the interaction mechanism probabilities increase with Z, particularly a photoelectric absorption which is required for a full energy deposition. For electrons, high Z materials tend to backscatter the electrons and promote bremshralung, both of which are undesirable.

# Semiconductor Diode problems

## ■ Problem 11.1. Intrinsic and doped concentrations.

First, from Table 11.1 we take the necessary constants for Si and Ge (i.e. the intrinsic carrier densities, or "$n_i$," the atomic weights, or "$A$," and the densities, or "$\rho$"). These constants are shown below.

$$n_{iSi} = \frac{1.5 \times 10^{10}}{cm^3} \qquad A_{Si} = \frac{28.09 \text{ grams}}{mole} \qquad \rho_{Si} = \frac{2.33 \text{ grams}}{cm^3}$$

$$n_{iGe} = \frac{2.4 \times 10^{13}}{cm^3} \qquad A_{Ge} = \frac{72.6 \text{ grams}}{mole} \qquad \rho_{Ge} = \frac{5.32 \text{ grams}}{cm^3}$$

We know that the material is "doped" if $n_d \gg n_i$ ($n_d$ is the dopant concentration). For Si or Ge, this begins at, say, $n_d \sim n_i$. What atom fraction is this in parts per billion? First, we answer this question for Si. To calculate this, we take $n_i$ and divide it by $\rho$ to give units of $\frac{carriers}{gram}$, then multiply that by A to give units of $\frac{carriers}{mole}$, then divide that by Avagadro's number to give units of $\frac{carriers}{atom}$, or $\frac{impurity\ atoms}{Si\ atoms}$, which is the atom fraction we are interested in. This calculation is shown below. The approximate atom fraction (in parts per billion) of impurity levels in Si such that we consider it to be "doped" is thus:

$$\text{atom fraction}_{Si} = \frac{n_{iSi}\,A_{Si}}{\rho_{Si}(\text{Avogadro's Constant})} = 0.0003003 \text{ parts per billiion}$$

Now, we do exactly the same calculation for Ge. The approximate atom fraction (in parts per billion) of impurity levels in Ge such that we consider it to be "doped" is thus:

$$\text{atom fraction}_{Ge} = \frac{n_{iGe}\,A_{Ge}}{\rho_{Ge}(\text{Avogadro's Constant})} = 0.544 \text{ parts per billion}$$

Note what this says. Because Ge has a much higher intrinsic carrier density (x 1000), we require a doping concentration in Ge 1000 times larger than in Si in order to have a "doped" material.

## ■ Problem 11.3. Expected electron-hole pairs in Si by 100 keV energy deposition.

The mean value of the number of electron-hole pairs (carriers) produced is the energy deposition divided by the energy required to produce a single electron-hole pair. This calculation is shown below. The mean number of electron-hole pairs produced is thus:

$$<\text{carriers}> = \frac{100 \text{ keV}}{\frac{3.76 \text{ eV}}{carrier}} = 26\,600 \text{ carriers}$$

The variance is given by the statistical variance (i.e. the mean) times the Fano factor (we use an approximate value F=0.1). This calculation is shown below:

$$\text{variance} = 0.1 \times 26\,595.7 \text{ carriers} = 2660 \text{ carriers}$$

**This is a brief review of the key relationships dealing with depleted semiconductors:**

Depletion depth : $d(V, \epsilon, N) = \sqrt{\dfrac{2\,\epsilon\,V}{\text{(Electron Charge)}\,N}}$ which can also be written as : $d(V, \epsilon, \mu, \rho_d) = \sqrt{2\,\epsilon\,V\,\mu\,\rho_d}$

Capacitance per unit area : $C(\epsilon, N, V) = \sqrt{\dfrac{\text{(Electron Charge)}\,\epsilon\,N}{2\,V}}$

Maximum Electric Field Strength at Junction : $E_{\max}(V, d) = \dfrac{2\,V}{d}$

- # Problem 11.5. Which bias on which contact?

For a reverse bias, one wants to extract the majority carrier by applying the opposite polarity. So, the positive voltage is applied to the n+ contact. This removes the excess electrons, and leave the fixed positive charge at the positive voltage. This positive voltage pulls on the electrons in the opposite p+ contact where they are only a minority carrier.

- # Problem 11.7. Largest depletion depth per unit voltage

The depletion depth is given by $\sqrt{\dfrac{2\,\epsilon V}{eN}}$ where N is the lowest dopant concentration. By making N as low as possible, i.e., by starting with the purest of materials, one achieves the highest depletion depth for any given voltage.

- # Problem 11.9. Energy loss measurement in dead layer.

To solve this, we look at the two equations that describe the two situations we have. Each equation describes the energy loss of the alpha particles in silicon when the alpha particles are either perpendicular or at an angle to the surface of the detector. We know that the energy lost is $\Delta E = E_i - E_f = E_a - hc$, where $E_a$ is the incident alpha particle energy, h is the channel number of the alpha peak (depends on the situation), and c is the energy per channel number of the MCA (we assume a zero offset, so this is valid). For the case of the alpha particles perpendicular to the detector, we have $\Delta E_0 = t\dfrac{dE_0}{dx}$ (Eqn. 11.22), so then we can say $E_a = t\dfrac{dE_0}{dx} + h_0 c$ for the case of perpendicular alphas ($h_0$ is the MCA channel number corresponding to this situation). For the case of the alpha particles at an angle to the detector, we have $\Delta E(\theta) = \dfrac{\Delta E_0}{\cos(\theta)}$ (Eqn. 11.23) $= \dfrac{t\dfrac{dE_0}{dx}}{\cos(\theta)}$, so then we can say $E_a = \dfrac{t\dfrac{dE_0}{dx}}{\cos(\theta)} + h_p c$ for the case of alphas at an angle ($h_p$ is the MCA channel number corresponding to this situation). These equations for $E_a$ are the equations that we would like to solve simultaneously for the dead layer thickness, t. We express these two equations below, where we denote $\dfrac{dE_0}{dx}$ with "$c_1$" and c with "$c_2$," and we solve for t by eliminating $c_2$, which is unknown.

$$E_a = t\,c_1 + h_0\,c_2 \qquad\qquad E_a = \dfrac{t\,c_1}{\cos(\theta)} + h_p\,c_2$$

The solution for t in terms of our known variables.

$$t = -\frac{E_a\, h_p - E_a\, h_0}{h_0\, c_1 \sec(\theta) - h_p\, c_1}$$

Here, we just plug our known values into the solution for t. Note that we get $E_a$ to be roughly 5.486 MeV (from Table 1.3 for the $E_\alpha$ of $Am^{241}$ in the dominant decay branch), and also, we approximate $\frac{dE}{dx}$ (i.e. $\frac{dE_0}{dx} = c_1$) in Si at this energy to be 120 keV/micron (from Figure 2.9). We substitute $E_a = 5.486$ MeV, $h_p = 455$, $h_0 = 461$, $c_1 = \frac{120\,keV}{\mu m}$ and $\theta = 35°$ to get the dead layer thickness in units of microns.

$$t = 2.55 \ \mu m$$

We are asked for the dead layer thickness in units of alpha particle energy loss, i.e. how much alpha particle energy is lost when the alphas are perpendicular to the detector surface? This is simply calculated by $t\frac{dE_0}{dx} = t * c_1$, which is shown in the below. The dead layer thickness of the silicon junction detector in units of (perpendicular) alpha particle energy loss (in keV) is thus:

$$t = 2.55 \ \mu m \ \times \ \frac{120\,keV}{\mu m} = 305\,keV$$

### ■ Problem 11.11. Electron and hole collection times in Si.

The maximum collection time is given by $t_c = \frac{d}{v}$, where we assume the particles must drift across the entire wafer (so d = 0.1 mm = 0.01 cm). From Figure 11.2 (a and b), the saturated electron and hole velocities (v) both appear to be about $10^7$ cm/s. The calculation for $t_c$ is shown below.

$$t_c = \frac{d}{v}$$

We substitute d=0.01 and v=$10^7$ to get the approximate maximum collection time for both electrons and holes in seconds.

$$t_c = 1 \times 10^{-9} \text{ seconds}$$

This very fast collection time is characteristic of thin semiconductor detectors.

### ■ Problem 11.13. Radiation damage to Si.

The text notes that serious damage occurs by $10^{11} \ \alpha/cm^2$. Given a 10 MBq $\alpha$ source at 10 cm from the detector, the flux (in $\frac{\alpha}{cm^2 * s}$) is $\frac{S\frac{\Omega}{4\pi}}{A}$, where $\Omega \sim A/d^2$. Hence, to find the length of exposure time required for damage to occur, we simply take the flux multiplied by the unknown time, set it equal to $10^{11} \ \alpha/cm^2$, and solve for the time. This calculation is expressed below.

$$\frac{S\,t}{4\pi d^2} = 10^{11}\,cm^{-2}$$

We substitute S=10 MBq, and d=10 cm to get the length of exposure time required for radiation damage to occur to the Si surface barrier detector.

$$t = 4\,000\,000\,\pi \text{ seconds} = 145 \text{ days}$$

This says that the detector will be seriously damaged if left in this position for ~5 months.

■ **Problem 11.15. Pulse height spectra for alphas for different depletion depths.**

(a). In this case, all of the energy is deposited and the pulse height is of corresponding amplitude.

(b). Since alphas lose most of their energy at the end of their path, less than half of the energy is deposited in the detector, so the pulse amplitude is less than half of that in part (a).

(c). From (b) above, the energy deposited must be more than half of that in part (a).

## Germanium gamma-ray detector problems.

### ■ Problem 12.1. Fano factor role.

Decreasing the Fano factor F plays a major role in adjusting the FWHM and, hence, the energy resolution downward. Recall from Eqn. 4.15 that

FWHM $\propto 2.35 \sqrt{N} \sqrt{F}$ , and the energy resolution is given by: $2.35 \sqrt{\dfrac{F}{N}}$ . If the Fano factor decreases by a factor of 2, both the FWHM and energy resolution are decreased by $\sqrt{2}$ .

### ■ Problem 12.3. Resolution contributions: statistics, electronics, incomplete charge collection negligible.

We look at the component representing the statistical fluctuation in the number of charge carriers from Eqn. 12.12, $W_D$, which is calculated by

FWHM $= 2.35 K \sqrt{N} \sqrt{F} = 2.35 \, \epsilon \sqrt{F * \dfrac{E}{\epsilon}}$ $(= 2.35 \sqrt{F \epsilon E}$ , from Eqn. 12.13; either equation works just as well), where $\epsilon$ is the energy/charge carrier (ionization energy), and E is the total energy deposited in the detector (140 keV). We use Fano= 0.08 and $\epsilon$ = 2.96 eV/carrier. The calculation for $W_D$ is shown below.

$$W_D = 2.35 \sqrt{\text{Fano} \times 2.96 \text{ eV} \times 140 \text{ keV}} \quad = 428 \text{ eV}$$

Note that this says that the statistical component, .427 keV, is much smaller than the electronic noise contribution of 1.2 keV (which will dominate). This is a realistic problem, and one is always looking for low noise electronic components which make the carrier statistics the dominant factor. Since the electronic noise is constant but the carrier statistics vary with $\sqrt{N}$ , the electronic component is especially critical for low energy applications, such as in nuclear medicine where 140 keV is used extensively. Frequently, people will quote the electronic noise in electrons -- this is just a factor of 2.96 eV (about 3 eV) to convert from energy to electrons -- about 140 electrons in our example above.

Below, we use Eqn. 12.12 to calculate $W_T$ as a percentage of the total energy deposited (note that we have assumed $W_X = 0$, i.e. we are assuming charge collection is complete).

$$\frac{\sqrt{(1.2 \text{ keV})^2 + W_D^{\,2}}}{140 \text{ keV}} = 0.910 \%$$

This is a typical (i.e., remarkably good) value for the energy resolution of a semiconductor detector.

### ■ Problem 12.5. Escape peaks in Ge versus NaI.

Ge detectors are typically smaller in size and have lower value of Z. Both of these factors lead to increased probability of escape for the 511 keV annihilation photons.

■ **Problem 12.7. Expected energy resolution for Ge at 662 keV.**

Since we can assume charge collection to be complete and electronic noise negligible, then the energy resolution is only a function of charge carrier statistics. Recall that the FWHM from charge carrier statistics (or $W_D$) is given by $2.35 \sqrt{F\epsilon E}$, where $\epsilon = 2.96$ eV/carrier-pair, F=0.08 (both from pg. 368), and E = 662 keV. This calculation for the FWHM is shown below.

$$\text{FWHM} = 2.35 \sqrt{F \epsilon E}$$

We substitute $\epsilon = 2.96$ eV, F = 0.08, E = 662 keV to get the FWHM in eV.

$$\text{FWHM} = 930 \text{ eV}$$

Now, to calculate the energy resolution, we simply divide the FWHM by the gamma-ray energy. This is shown below as a percent.

$$\frac{930.439 \text{ eV}}{662 \text{ keV}} = 0.141 \text{ \%}$$

## Other Semiconductor Devices.

### ■ Problem 13.1. X-ray escape peak smaller for Si than Ge.

Two reasons the x-ray escape peak from Si is smaller are:

(1). Greater penetration of the incident radiation into Si rather than Ge (smaller photoelectric cross section in Si) leads to fewer events occurring near the surface in Si, which is where the x-ray has the greatest probability of escape.

(2). A greater x-ray energy for Ge (11 keV) as opposed to Si (1.8 keV) allows the Ge x-rays to escape more easily (higher probability of escape).

### ■ Problem 13.3. Required resolution to separate Cu and Zn $K_\alpha$ X-rays.

The closest x-rays of interest are Cu $K_{\alpha 1}$ at 8.048 keV and Zn $K_{\alpha 2}$ at 8.616 keV (using these x-ray energies will give us the resolution necessary to resolve separately any of the K-characteristic x-rays from Cu and Zn). We want the FWHM to be (at least) equal to the difference between these two energies, $\Delta E$, to resolve separately the X-rays. We know that $R = \dfrac{\text{FWHM}}{E}$, or here, $R = \dfrac{\Delta E}{E_{avg}}$. The calculation for the required energy resolution is shown below.

$$R = \frac{\Delta E}{E_{avg}} = \frac{8.616 - 8.048}{\frac{8.616 + 8.048}{2}}$$

The energy resolution expressed as a fraction, not a percent is thus:

$$R = 0.0682$$

So, we need an energy resolution of about 6.8% to resolve separately the K-characteristic x-rays from copper and zinc.

### ■ Problem 13.5. Efficiency of planar Si(Li) detector versus NaI well-counter.

This problem is solved using the equation $N = S\, \epsilon_{abs,p} = S\, \epsilon_{ip}\dfrac{\Omega}{4\pi}$, and the intensity ratio (X/$\gamma$ from Table 13.2, which is the same as $\dfrac{S_X}{S_\gamma}$) for Co-57 at an X-ray energy of 6.4 keV (or 6.397 keV), which is given as 0.5863. In the end, we are interested in finding $\epsilon_{ip}$ for this x-ray energy in Si(Li) (denoted as $\epsilon_{ip,X}$). We start by noting that $S_\gamma = \dfrac{N_\gamma}{\epsilon_{abs,p,\gamma}}$ ($\epsilon_{abs,p}$ and $N_\gamma$ are both given for the gamma-rays in NaI(Tl)), and then we note that $S_X = (\dfrac{S_X}{S_\gamma})S_\gamma = .5863\, S_\gamma = .5863\left(\dfrac{N_\gamma}{\epsilon_{abs,p,\gamma}}\right)$. From this, we can calculate $\epsilon_{abs,p}$ for the x-rays in Si(Li) by $\epsilon_{abs,p,X} = \dfrac{N_X}{S_X} = \dfrac{N_X}{.5863\left(\dfrac{N_\gamma}{\epsilon_{abs,p,\gamma}}\right)} = \epsilon_{abs,p,\gamma}\left(\dfrac{N_X}{.5863\, N_\gamma}\right)$. The calculation for $\epsilon_{abs,p,X}$ is shown below.

$$\epsilon_{abs,p,X} = .83\left(\frac{730}{60}\right)\left(\frac{15}{.5863 \times 146\,835}\right)$$

The absolute peak efficiency for the Co-57 X-rays (expressed as a fraction, not a percent) is thus:

$$\epsilon_{abs,p,X} = 0.00176$$

From this, we calculate $\epsilon_{ip,X}$ from $\epsilon_{ip}= \dfrac{\epsilon_{abs,p}}{\left(\dfrac{\Omega}{4\pi}\right)}$, and we approximate $\Omega$ as $\dfrac{A}{d^2}$ (A is expressed in cm² since d is in cm). This calcula-

tion is expressed below.

$$\epsilon_{ip,X} = \dfrac{\epsilon_{abs,p,X}}{\dfrac{\Omega}{4\pi}}$$

We substitute $\Omega = \dfrac{A}{d^2}$, A = 3 and d = 10 to get the intrinsic peak efficiency of the Si(Li) detector for the 6.4 keV X-rays from Co-57 (expressed as a fraction, not a percent).

$$\epsilon_{ip,X} = 0.737$$

This high intrinsic peak efficiency is comparable to the NaI(Tl) well counter. Note how the geometric factor of moving the source just 10 cm away reduces the efficiency by about $10^3$!

- **Problem 13.7. Detector thicknesses needed: Si, Ge, CdTe, and HgI$_2$ are compared.**

This problem shows the advantages of higher Z (i.e. higher attenuation coefficients) in requiring smaller quantities of material for equivalent detection efficiencies. The probability of having at least one interaction in thickness t is given by $1 - e^{-\mu t}$ (since the probability of having zero interactions in thickness t is $e^{-\mu t}$, same as $\dfrac{I}{I_0}$). We want this probability to be 50%, and we are interested in finding t, so we set this equation equal to 0.5 and solve for t in terms of $\mu$:

$$1 - e^{-\mu t} = \dfrac{1}{2}$$

The solution for t in terms of $\mu$.

$$t = \dfrac{\ln(2)}{\mu}$$

The only data needed are the attenuation coefficients for {Si, Ge, CdTe, HgI$_2$}, found in the textbook for incident 662 keV gamma-rays. These are expressed below in cm$^{-1}$, where we have one set for photoelectric absorption, one set for Compton scattering, and one set that sums the two to give the total attenuation coefficients (662 keV is too low of an energy for pair production, so the attenuation coefficients for pair production would be zero).

$$\mu_{pe} = \left\{ \dfrac{2.13}{10^4}, \dfrac{9.48}{10^3}, \dfrac{5.08}{10^2}, \dfrac{1.50}{10} \right\}$$

$$\mu_C = \left\{ \dfrac{1.78}{10}, \dfrac{3.60}{10}, \dfrac{3.86}{10}, \dfrac{3.99}{10} \right\}$$

$$\mu_T = \mu_C + \mu_{pe}$$

Adding the photoelectric and Compton attenuation coefficients gives the total attenuation coefficient, $\mu_T$. We list the values of $\mu_T$, (in cm$^{-1}$) for {Si, Ge, CdTe, HgI$_2$}:

$$\mu_T = \{0.178, 0.369, 0.437, 0.549\}$$

We plug these attenuation coefficients into our solution for t to give the required thicknesses for each detector.

$$t = \frac{\ln(2)}{\mu_T}$$

The required thicknesses (in cm) for Si, Ge, CdTe, and HgI$_2$, respectively, such that the probability of a 662 keV gamma-ray having at least one interaction is 50% is thus:

$$t = \{3.89,\ 1.87,\ 1.59,\ 1.26\}\ \text{cm}$$

Note the remarkable advantage of high Z, high density materials like HgI$_2$. One gains a factor of 2-3 (in each dimension), so the required volumes are substantially reduced.

The other question asks for the fraction of interactions at 662 keV that are photoelectric events (desired for high photopeak efficiencies). This is simply the photoelectric attenuation coefficients divided by the total attenuation coefficients:

$$f_{pe} = \frac{\mu_{pe}}{\mu_T}$$

The fraction of interactions at 662 keV that are photoelectric events for Si, Ge, CdTe, and HgI$_2$, respectively:

$$f_{pe} = \frac{\mu_{pe}}{\mu_T} = \{0.0012,\ 0.0257,\ 0.116,\ 0.273\}$$

Again, note the distinct advantage of the high Z materials -- silicon and germanium have a very low percentage of single interaction photoelectric events, while the higher Z materials of CdTe and HgI$_2$ do dramatically better. This also implies much smaller volumes required to acquire full energy depositions.

# Slow neutron detectors

## ▪ Problem 14.1. Operation in Proportional Region

We choose to operate gas-filled neutron detectors in the proportional region. If we operated in the ionization region, the individual pulses would be inconveniently small to handle, as noted in problem 14.2 below. In the Geiger region, any interaction in the detector would yield the same size pulse, so we would be unable to discriminate based on pulse height against the small pulses produced by gamma-ray interactions, background, or noise sources.

## ▪ Problem 14.3. Gain in efficiency using enriched $^{10}B$.

Natural boron is composed of 19.8% $^{10}B$. $^9B$ has a negligible cross section for $(n, \alpha)$. First, using Eqn. 14.4, we find solve for $f = \Sigma_a(E)\,L$ when the detection efficiency is 1% (i.e. $^{10}B$ enrichment = 19.8%). This is done below, where we set Eqn. 14.4 equal to .01 (with the appropriate "f" substitution made) and solve for f.

$$1 - e^{-f} = .01$$

The quantity $f = \Sigma_a(E)\,L$ for $^{10}B$ enrichment = 19.8% (efficiency = 1%) is thus:

$$f = 0.0101$$

Recall that $\Sigma_a(E) = N_{B-10}\,\sigma_{(n,\alpha)}$, and so is linearly proportional to the $^{10}B$ enrichment through the number density $N_{B-10}$. Now, to find the gain in efficiency when the $^{10}B$ enrichment is raised to 96%, we must divide the efficiency at this enrichment with the efficiency at a $^{10}B$ enrichment of 19.8% (efficiency of 1%). To find the new efficiency, we note that $\dfrac{f_{new}}{f_{old}} =$

$\dfrac{\Sigma_a(E)_{new}}{\Sigma_a(E)_{old}}$ (since L does not change) $= \dfrac{\text{new }^{10}B \text{ enrichment}}{\text{old }^{10}B \text{ enrichment}}$ (since the neutron energy does not change). Therefore, the quantity "f" that we use for the new efficiency is simply f calculated above (or $f_{old}$) multiplied by the ratio of the new $^{10}B$ enrichment to the old $^{10}B$ enrichment. The calculation of this efficiency gain is expressed below.

$$\text{efficiency gain factor} = \frac{1 - e^{-\frac{f\,.96}{.198}}}{.01}$$

We substitute f=0.0101 to get the detection efficiency gain for 10 eV neutrons when natural boron is replaced with boron enriched to 96% $^{10}B$ in a $BF_3$ tube:

> **efficiency gain factor = 4.76**

Note that this is close to the ratio of the abundances ($\dfrac{96\,\%}{19.8\,\%} = 4.85$) because of the relatively low attenuation (i.e., in this case, $e^{-f} \sim$ 1-f).

■ **Problem 14.5. BF₃versus ³He**

He-3 has the advantages of being easy to work with, being nontoxic and noncorrosive, and has the ability to be used at quite high pressures. Therefore, when efficiency is the primary concern, He-3 is preferred. In general, it is the preferred gas of choice for slow neutron detection, and most manufacturers choose this material in modern thermal neutron detectors. Unfortunately, because the source of He-3 comes from the diminishing weapons program, there is a forecast shortage of this isotope, and researchers are looking at viable alternatives or nontradiational sources to supply the demand for He-3.

■ **Problem 14.7. Ultra-thin-wall boron-lined detector spectrum.**

If there is no energy loss in the wall, then the two possibilities are either a full energy deposition of the alpha particle (1.47 MeV) if the alpha particle is ejected into the gas or a full energy deposition of the Li particle (0.84 MeV). Thus, we would expect to isolated peaks at these two energies.

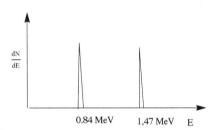

■ **Problem 14.9. Fission chamber efficiency limit.**

Increasing the thickness of the fissile material deposit increases the probability that an incidnet neutron will undergo fission, but for detection, at least one of the fission fragments must reach the active gas to cause an ionization. Thus, any fissile material located further from the inner wall than the largest fission product range will not contribute to detection efficiency.

■ **Problem 14.11. Fractional burnup of Rhodium SPN detector.**

The number of reactions that have occurred after time t (number of target atoms lost) is approximated by $N_0\sigma\phi t$ (where $N_0$ is the original number of target atoms ... we ignore the slight loss due to burnup), so the fractional burnup is just $\sigma\phi t$. The thermal absorption cross section is given as $\sigma_{th} = 150$ barns ( we sum the Table 14.1 two values of 139 and 11 barns), the flux is given as $3 \times 10^{13} /(cm^2 * s)$, and the exposure time is given as 6 months. We use this equation to approximate the fractional burn–up after 6 months of exposure:

$$\text{fractional burn–up} = \frac{(150 \text{ barns}) \left(3 \times 10^{13}\right) (6 \text{ months})}{cm^2 \text{ second}} = 0.0710$$

A more rigorous approach is to write the number of target atoms as N and note $\frac{dN}{dt} = -N\sigma\phi$ (by assuming a steady-state current), so that $\frac{N(t)}{N_0} = e^{-\sigma\phi t}$. The fractional burn-up is given by $\frac{(N_0 - N(t))}{N_0} = 1 - \frac{N(t)}{N_o} = 1 - e^{-\sigma\phi t}$ (which we had approximated as $\sigma\phi t$ above). We calculate the fractional burn-up using this more accurate equation below:

$$\text{fractional burn--up} = 1 - e^{-\frac{(150\ \text{barns})\,(3\ x\ 10^{13})\,(6\ \text{months})}{\text{cm}^2\ \text{second}}} = 0.0685$$

where the unit conversions of barns-to-$\text{cm}^2$ and months-to-seconds are not shown in the equation above. Thus, the sensitivity of the detector drops by about 7% over six months. In the case where one is monitoring reactor power using the SPND, this must be accounted for to avoid underestimating the true neutron flux in the core.

# Fast Neutron detectors

## ▪ Problem 15.1. Efficiency of $^6$LiI scintillator for 1 MeV neutrons.

We know that $\epsilon(E) = 1-\exp[-\Sigma_a(E) L] = 1-\exp[-N_{LiI} \sigma_a(E) L]$. For simplicity, we assume that 100% enrichment of $^6$Li and we neglect absorption of neutrons in I. From Fig. 14.1, we find that $\sigma_a$(1 MeV)= .25 barns, and from pg. 522 we know that $\sigma_a$(.025 eV)= 940 barns. We are also given that t (=L) = 4 mm. Therefore, the first task is to calculate the atom density of LiI, $N_{Li}$. This is done by noting that $\rho_{LiI}$= 4.08 $g/cm^3$(cf. Table 8.3), and the molar mass of LiI is 133 g/mole (=6+127). From this, it is easy to calculate the atom density:

$$N_{Li} = \frac{\dfrac{4.08\,g}{\text{mole}} \times (\text{Avogadro's Constant})}{\dfrac{133\,g}{\text{mole}}} = \frac{1.84739 \times 10^{22}}{cm^3}$$

Given $N_{Li}$, we calculate $\epsilon(E)$ below.

$$\epsilon(E) = 1 - e^{N_{Li}\sigma_a t}$$

For fast neutrons, we substitute $\sigma_a = $ .25 barns and $t = $ 4 mm and the calculated value of $\epsilon(E)$ is :

$$\epsilon_{fast}(E) = 0.00185$$

For thermal neutrons, we substitute $\sigma_a = $ 940 barns and $t = $ 4 mm and the calculated value of $\epsilon(E)$ is :

$$\epsilon_{thermal}(E) = 0.999$$

We learn that the efficiency for 1 MeV neutrons is ~500 times smaller than the efficiency for thermal neutrons. This means that the detector will usually have a peak, i.e. an epithermal peak, even when we are trying to measure predominately fast neutrons since thermal neutrons are ubiquitous. Note that even a 4 mm thick detector is essentially black (i.e., efficiency ~ 100%) to thermal neutrons.

## ▪ Problem 15.3. LiI detector

At the center of the moderating sphere is a LiI detector. Since these are now thermal neutrons that are to be detected, the incident neutrons bring a neglibly small incident energy. The most likely ensuing reaction is thus $^6_3Li(n, \alpha)\, ^3_1H$ with a Q value of 4.78 MeV. Thus, we would expect to see a single photopeak at 4.78 MeV since both reaction products will likely be contained in the detector.

## ▪ Problem 15.5. Find $E_{alpha}$ and $E_{triton}$ in Li sandwich spectrometer. $E_n$ = 3 MeV.

This is similar to Problem 1.3. We must satisfy conservation of momentum and energy. The variables are self-explanatory, and the conservation of momentum and energy are written as:

$$p_n = p_\alpha + p_t \qquad KE_n + Q = KE_\alpha + KE_t \qquad KE_n = \frac{p_n^2}{2\,m_n}$$

$$KE_\alpha = \frac{p_\alpha^2}{2\,m_\alpha} \qquad KE_t = \frac{p_t^2}{2\,m_t} \qquad m_t = 3\,m_n \qquad m_\alpha = 4\,m_n$$

We solve these quadratic equations for $KE_\alpha$ and $KE_t$ (all energies are in MeV) by eliminating momenta, and find the reaction product kinetic energies:

$$KE_\alpha = \frac{1}{49}\left(-4\sqrt{3}\,\sqrt{7\,Q\,KE_n + 6\,KE_n^2} + 22\,KE_n + 21\,Q\right)$$

$$KE_t = \frac{1}{49}\left(4\sqrt{3}\,\sqrt{7\,Q\,KE_n + 6\,KE_n^2} + 27\,KE_n + 28\,Q\right)$$

and

$$KE_\alpha = \frac{1}{49}\sqrt{3}\,4\sqrt{7\,Q\,KE_n + 6\,KE_n^2} + \frac{22\,KE_n}{49} + \frac{3\,Q}{7}$$

$$KE_t = \frac{1}{49}\left(-4\sqrt{3}\,\sqrt{7\,Q\,KE_n + 6\,KE_n^2} + 27\,KE_n + 28\,Q\right)$$

We note that there are two possible solutions for $KE_\alpha$ and two solutions for $KE_t$. We substitute Q=4.78 and $KE_n = 3$ to evaluate both of them.

1)  $KE_\alpha = 6.14$    $KE_t = 1.64$

2)  $KE_\alpha = 2.63$    $KE_t = 5.15$

The two solutions correspond to the cases where the alpha goes in the forward direction or backward direction, respectively. The former is the case for this problem. The fact that both particles have > 1 MeV worth of energy is encouraging. We could take these energy values, and look at the particles' ranges in Li to get an idea of the maximum thickness of Li we can have and still have the reaction products emerge into the Si diode detector for detection.

## ▪ Problem 15.7. Thermal peak for fast neutrons

Although we may be trying to measure the fast neutron spectrum, in practice there will always be thermal neutrons also present. This is inevitable since the surrounding environment, such as the walls, floor, and/or people, will serve as a moderating material that will return some thermal neutrons to the detector. These will give rise to a detected peak at the Q value of the detection reaction, which, for He-3, will be 764 keV.

## ▪ Problem 15.9. Fast Neutron Scatterings separated by 3 cm.

We recall that, in the laboratory frame of reference, $E_n' = E_n \cos^2 \theta_n$. (You should be able to prove this using the conservation of momentum and energy equations). The time between the two scatterings is given by $t = \Delta x / \sqrt{2 E_n'/m}$ ] so:

$$t = \frac{\Delta x}{\sqrt{\frac{2\,KE_n \cos^2(\theta)}{m_n}}}$$

where we substitute $\theta = 40°$, $KE_n = $ MeV and $\Delta x = 3$ cm to yield :

$t = 2.83$ nano seconds

This time is significantly less than the anode time constant, so the two events cannot be resolved in time.

### Problem 15.11. Show the angle between scattered neutron and reoil proton must be 90 degrees in lab frame.

This is easiest to prove using vector notation. Writing conservation of momentum in vector notation: $\overline{p}_p + \overline{p}_{n'} = \overline{p}_{n_o}$.

Square this equation (i.e., dot this equation with itself) and then using $m_n = m_p = m$, and $p^2 = 2\,mE$, and applying conservation of energy yields:

$\overline{p}_p \bullet \overline{p}_{n'} = 0.$

This can only be always true if the angle between these two vectors is 90 degrees.

### ■ Problem 15.13. Si recoil energies from 1 MeV neutrons

The recoil nucleus energy can vary from 0 (for a glancing 90° scatter), up to a direct hit (0°) yielding a maximum Si recoil energy of:

$$\text{recoil energy} = \left( \frac{4\,A}{(A + 1)^2} \right) KE_n \qquad \text{where we substitute } A = 28 \text{ and } KE_n = 1 \text{ MeV}$$

$$\text{recoil energy} = 133 \text{ keV}$$

Note that this is very small -- only 133 keV -- and it won't travel very far at all since it is massive and highly charged, but it is large enough to dislodge the atom from its place in the lattice. These crystal defects can lead to degraded energy resolution over time, and is the reason that semiconductor detectors are often avoided when fast neutrons must be measured. This is a particular problem for long space flights, for example, and detectors are sometimes annealed during the transit.

### ■ Problem 15.15. Efficiency of proton radiator in proton recoil spectrometer.

The proton recoil spectrometer can provide an accurate measurement of the fast neutron spectrum, but is usually quite inefficient. The efficiency can be improved if the hydrogenated scattering material is made thicker, but only up to a point. The detection reaction is (n,p), and it is the proton which must escape from the radiator with a neglible loss of energy and then its energy must be accurately detected separately. If the scattering material becomes too thick, the proton will not be able to escape without losing a significant fraction of its kinetic energy.

### ■ Supplemental Problem: Fast neutron recoil spectrometer. Statistics of Distributions

Two hydrogenous scintillators of identical dimensions are fabricated for use as neutron recoil spectrometers. While the first detector is composed of only ordinary protons, the second detector has only deuterons. Assuming that the scattering cross sections are identical for protons and deuterons,

(a). graph the spectra expected from monoenergetic fast neutrons of En = 2.5 MeV using: (i) the protonated scintillator, (ii) the deuterated scintillator, and (iii) the subtracted spectrum, i.e., (spectrum (i) - spectrum (ii))

(b). Suppose that instead of waiting for the two spectra to be completed and then subtracting them, we measure an energy from the protonated scintillator, subtract the energy of the next event measured by the deuterated scintillator, and then record the net energy measured. In this case, what does the expected differential pulse height spectrum look like?

# Pulse processing and shaping

## ▪ Problem 16.1. Pulse travel time

For the cable RG-59U, Table 16.1 lists the pulse speed as .659c, so to travel d=15 meters would take:

$$t = \frac{d}{v} = 7.59 \times 10^{-8} \text{ seconds}$$

## ▪ Problem 16.3. Situations for which cable termination is needed

Terminating cables in their characteristic impedance is important to avoid reflections of pulses and to transmit the full pulse amplitude to subsequent signal processing elements. This is most critical for pulses whose rise time is short compared to the pulse transit time.

(a). Signals in RG-59/U cables travel at 0.659c, so in 20 m, the pulse transit time is:

$$t_{\text{transit}} = \frac{20}{0.659 \times 3 \times 10^8} = 101 \text{ ns}$$

If the pulse has a rise time of 500 ns, then because this is long compared to the transit time (i.e., the pulse is "slow"), proper cable termination is less important.

(b). Signals in RG-62U cables travel at 0.84c, so in 10 m, the pulse transit time is:

$$t_{\text{transit}} = \frac{10}{0.840 \times 3 \times 10^8} = 39.7 \text{ ns}$$

If the pulse has a rise time of 10 ns, then because the rise time is short compared to the transit time (i.e., the pulse is "fast"), matching the input impedance of the next device in the signal chain through the use of a terminating resistance is important.

## ▪ Problem 16.5. Pulse attenuator

We use a pulse attenuator shown in Figure 16.8(b). The key to this design is that the input impedance must be $R_0$ and the load also has an equivalent resistance of $R_0$. The equations shown in the Figure are derived below, where the first equation states that the equivalent input impedance (shown on the right side of the equation) is to be $R_0$.

$$R_0 = R_1 + \frac{R_2\,(R_1 + R_0)}{(R_1 + R_2 + R_0)} \qquad \alpha = \frac{V_{\text{in}}}{V_{\text{out}}} \qquad i_1 = i_2 + i_3$$

$$V_{\text{in}} = i_2\,R_2 + i_1\,R_1 \qquad V_{\text{in}} = i_3\,R_1 + i_3\,R_0 + i_1\,R_1 \qquad V_{\text{out}} = i_3\,R_0$$

Solving the above equations for $R_1$ and $R_2$ by eliminating $V_{\text{in}}$, $V_{\text{out}}$, $i_1$, $i_2$ and $i_3$ yields:

$$R_1 = \frac{R_0\,(\alpha - 1)}{\alpha + 1} \quad \text{and} \quad R_2 = \frac{2\,R_0\,\alpha}{(\alpha - 1)\,(\alpha + 1)}$$

For the problem being asked, we substitute $\alpha=10$ and $R_0 = 50\ \Omega$ to get our answers:

$$R_1 = 40.9 \, \Omega \quad \text{and} \quad R_2 = 10.1 \, \Omega$$

### ■ Problem 16.7. Decay time of exponential tail.

We are looking for the time at which the decayed pulse reaches only 1% of its original amplitude. At that time, another pulse riding on it will be measured to be too large by 1%. We use time units of microseconds and solve:

$$e^{-t/\tau} = .01$$

We solve the above equation for t and substitute $\tau=50$ to find the time at which the decayed pulse reaches only 1% of its original amplitude.

$$t = 230 \, \mu s$$

### ■ Problem 16.9. Charge sensitive pre-amplifier derivation.

The circuit equations of interest are (balancing charge flow and voltage drops):

$$V_{\text{in}} = \frac{Q_f}{C_f} + V_{\text{out}} \qquad V_{\text{in}} = \frac{Q_{\text{in}}}{C_{\text{in}}} \qquad V_{\text{out}} = -A \, V_{\text{in}} \qquad Q = Q_f + Q_{\text{in}}$$

Solving the system of equations above for $V_{\text{out}}$ gives us :

$$V_{\text{out}} = \frac{-A \, Q}{C_f + A \, C_f + C_{\text{in}}}$$

Note that the feedback resistor doesn't play a role here. All of the initial charge flow goes into charging the capacitors. Let's look at the solution more carefully by multiplying the numerator and denominator by $\frac{1}{A \, C_f}$ and expanding it out:

$$\text{Numerator :} \qquad -\frac{Q}{C_f}$$

$$\text{Denominator :} \qquad 1 + \frac{1}{A} + \frac{C_{\text{in}}}{A \, C_f}$$

If the denominator ~1, then $V_{\text{out}} \sim \frac{Q}{C_f}$. So we need $A \gg \frac{(C_{\text{in}} + C_f)}{C_f}$ for this condition to be true. ( To prove this, simplify the denominator:

$$1 + \frac{1 + \frac{C_{\text{in}}}{C_f}}{A}$$

For this denominator to be ~1, the condition above for A >> ... must be true).

# Pulse Shaping, Counting, and Timing

- **Problem 17.1. Find output form of large-RC integrating circuit with input voltage form $1\text{-}e^{-t/k}$.**

We write down the differential equation governing the time dependence of the output voltage pulse given an input shape of $1-e^{-t/k}$ , and solve it for the boundary condition:

$$v'(t) + \frac{v(t)}{\tau} = E_0\left(1 - e^{-t/k}\right) \qquad \text{where } v(0) = 0$$

which has the solution:

$$v(t) = \frac{E_0\,\tau}{(k-\tau)\,e^{t/\tau}} + \frac{E_0\left(\tau\,e^{t/k} + k\left(-e^{t/k}\right) + k\right)}{e^{t/k}\,(\tau - k)} = E_0\left(1 - e^{-\frac{t}{k}}\right)$$

This is the answer for the voltage as a function of time. To see what this looks like, let's choose $E_0=1$, $\tau=10$ and $k=1$. In this case RC is large, and we see integration and the passing of the low-frequency component. Try the example below with RC (i.e., $\tau$) small, and the output is the same shape as the input.

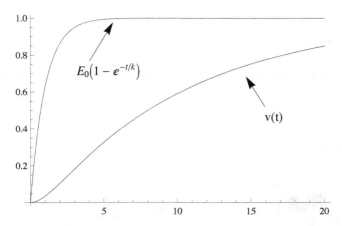

$$E_0\left(1 - e^{-t/k}\right)$$

$$v(t)$$

- **Problem 17.3 CR-RC network. Maximum amplitude of shaped pulse.**

The pulse shape is shown below, where we measure time in units of $\tau$:

$$y = t\,e^{-t}$$

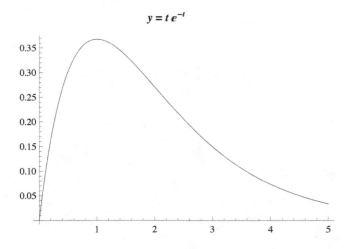

The maximum of the pulse occurs at the time where the derivative is zero:

$$\frac{d}{dt}\left(t\,e^{-t}\right) = 0$$

Taking the derivative, setting it equal to zero, and solving for t yields:

$$t_{max} = 1$$

and the value of the function there is:

$$y(1) = \frac{1}{e} = 0.368$$

so the maximum value is 37% of the input voltage and occurs at a time $t=\tau$.

### ■ Problem 17.5. Advantage of bipolar pulses

The bipolar pulse, because it has essentially 0 net voltage over its duration, is less susceptible to baseline shift. At high count rates, this is particularly advantageous.

### ■ Problem 17.7. Pulse shaping using a shorted delay line.

The idea is to take a monopolar pulse and combine it with the reflection from delay line to form a bipolar pulse. A diagram is useful.

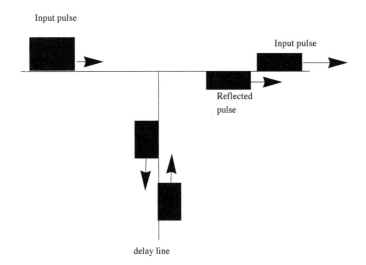

The time delay between the start of the transmitted input pulse and the start of the reflected pulse is just the transit time down and back the delay line. This time is $t = d/v = \dfrac{20}{0.659 \times 3 \times 10^8} = 101$ ns.

Since the input pulse is 200 ns wide, the reflected pulse meets up with it and cancels it after 101 ns (~100 ns). After 200 ns, the incident pulse has ended and the reflected pulse continues for 100 ns longer. The result is shown in the summation below:

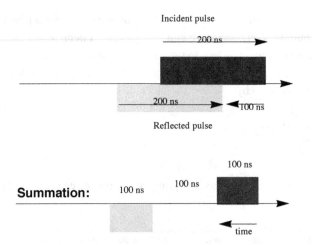

- **Problem 17.9. Rate Meter Time Constant Selection**

See text Chapter 17.II.E (p. 646). A longer time constant, RC, will smooth the output from random pulses so that the measured count rate is not jumpy. The disadvantage is the inability to catch rapid fluctuations in the count rate.

■ **Problem 17.11. Pileup losses.**

There is frequent confusion on the difference between deadtime losses and pileup losses. Deadtime losses are associated with pulse counting systems in which the system is incapable of counting the next event for a time $\tau$ (or longer) after the event. If $\tau$ is extendable, the system is termed paralyzable. For deadtime losses, we are concerned with the number of counts we get from the incoming events. The probability that an event will yield a count is $\frac{m}{n}$. Chapter 4 discusses this topic.

Having registered a count, we might also ask whether that count is pileup free. That is, that nothing occurs for a time $\tau$ after the arrival of the event being counted so that the measured amplitude is representative of a single event. Pileup losses are of concern with spectroscopy systems which must measure single event pulse amplitudes. In the case of multiple events following a count, we map the count to an improper channel. For a *count* to be pileup free, we must have no events arriving for a period $\tau$ after the count. So the *fraction of events* which give pileup free counts is given by:

Probability that an event yields a count • Probability that a count is pileup free

$$= \frac{m}{n} e^{-n\tau}.$$

Substituting for m/n for the nonparalyzable deadtime expression yields for this problem, the fraction of events yielding pileup free counts:

$$\text{pileup free counts} = \frac{e^{-n\tau}}{1 + n\tau} \quad \text{where we substitute } n = 25\,000 \text{ sec}^{-1} \text{ and } \tau = 4\times10^{-6} \text{ seconds to get our answer :}$$

$$\text{pileup free counts} = 0.823$$

■ **Problem 17.13. Bipolar shaping methods**

The double dealy line method and CR-RC-CR double differentiating methods both produce bipolar pulses. Review your textbook for the details of how these work.

■ **Problem 17.15. Output rate for an anticoincidence unit with coincident window $\tau$.**

The anticoincident unit will output a pulse only if there are NOT two pulses which arrive within a $\tau$ time of each other. The probability that a count will not arrive during time $\tau$ is given by $e^{-r\tau}$ so our answer is:

$$\text{probability} = r_1 e^{-r_2\tau} + r_2 e^{-r_1\tau}$$

For small values of $r\tau$, we expand the answer in a Taylor series around 0 to order 1:

$$\text{probability} = r_1 + r_2 - 2\tau r_1 r_2$$

This makes sense. We count $r_1 + r_2$ minus the chance coincident rate.

■ **Problem 17.17. Length of cable needed for 100 ns delay**

For RG-59 cable, the signal velocity is 0.659c or $1.98 \times 10^8$ m/sec or, inverting this, 5.06 ns/m. So for a delay of 100 ns, we require a length of cable, L:

$$L = \frac{100 \text{ ns}}{5.06 \text{ (ns}/m)} = 19.76 \text{ m}$$

■ **Problem 17.19. Improving true to chance coincidence ratio.**

In answering these questions, recall that the chance coincidence rate is $r_{true} = S \, \epsilon_1 \, \epsilon_2$ and $r_{chance} = (2 \, r_1 \, r_2 \, \tau) = 2 \, (S \, \epsilon_1 \, S \, \epsilon_2 \, \tau)$ so

$$\frac{r_{tr}}{r_{ch}} = \frac{1}{2 \, S \tau}$$

a. Changing the solid angle does not change S or $\tau$ so there is no effect.

b. Increasing S reduces the true to chance coincidence ratio as shown in the equation above.

c. Increasing $\tau$ reduces the ratio above.

d. Increasing the energy window slightly only changes the efficiency by which the pulses are selected. If one opens this window too large so that pulses other than the desired energy are included, then additional pulses that are not coincidence pulses are allowed, thereby effecively increasing the number of non-true coincidence pulses and thereby decreasing the ratio.

■ **Problem 17.21. Subranging ADC.**

A 12 bit-subranging ADC can consist of 3 stages of 4 bits or 4 stages of 3 bits. Which has the smallest (a) number of comparators, and (b) latency (assuming the same clock frequency)?

The *number of comparitors needed is related to the number of bits of information N according to* $2^N$ . So the 3-stages of four bit ADCs use $3 * 2^4 = 48$ comparators, whereas 4 stages of three bit ADCs use $4 * 2^3 = 32$ comparators. While the 3 stage ADC needs more comparators, the latency depends on the number of stages, however, so the *4-stage ADC can be expected to have a longer latency*. However, since the architecture involves serial*pipelining, the throughput should be proportional to the clock speed -- which we assumed was constant -- and should be the same for both cases.

Looking at the two extremes of the subranging ADC, an N-bit subranging ADC which has:
(a). 1 stage is known as a Flash ADC, and
(b). N stages is a successive approximation ADC.

## Multichannel Pulse Analysis

### ▪ Problem 18.1. Channels needed for fixed detector resolution

For a fixed resolution, we want the peak FWHM to contain 5 channels, so

FWHM = 5 channels = (Resolution)*(Peak centroid channel).

The ADC must have at least the Peak centroid channel number of channels, so the ADC must have:

$$ADC = \frac{5}{.003} = 1667 \text{ channels}$$

### ▪ Problem 18.3. Wilkinson ADC Oscillator Frequency.

We have to be able to resolve the maximum amplitude into 2048 parts in 25 $\mu$s. Each click of the clock corresponds to one channel number. The conversion time is given by $t_c = \frac{PH}{f}$ where the Pulse Height PH refers to the pulse height channel number. If the offset is zero, then:

$$f = \frac{PH}{t_c} = \frac{2048}{25 \, \mu s} = 81.9 \, \text{MHz.}$$

### ▪ Problem 18.5. Deadtime losses in Wilkinson ADC

If the ADC uses a linear ramp, then the deadtime is proportional to the pulse-height and not a fixed value. The expressions derived for the deadtime losses in Chapter 4 assumed that the deadtime was a fixed value, regardless of the pulse height. Note that other ADC designs, such as the Flash ADC or the Successive Approximation ADC have a fixed deadtime per conversion.

### ▪ Problem 18.7. Periodic pulser dead time.

The key to this problem is to note that the pulser pulses are periodic. Therefore, they will each be counted if the time between pulses is longer than the dead time per pulse ($t_{dead}$). Noting the time between the pulses of frequency f is $\Delta t = 1/f$, the count rate CR is less than f if $\Delta t > t_{dead}$ or $f < 1/t_{dead}$. If the frequency is high enough, multiple pulses may be included in the dead time. You can show that the count rate CR is given by:

CR= f/(n+1)   if   f > $(n/t_{dead})$ for n= {0,1,2, ....}

A plot of CR versus f (in units of $1/t_{dead}$) thus looks like a set of steps starting at height f and going down by half its height each step along the frequency axis.

### ▪ Problem 18.9. Channel counting statistics.

The average count in a channel should be zero, but this will fluctuate due to counting statistics. Indeed, half of the channels will have negative values. Since:  N = (S+B)-(B) = B - B, then by the propagation of error:

$\sigma_N$= Sqrt [2B] =

$$\sigma_N = \sqrt{600} = 24.5 \text{ counts}$$

So the average channel will have a value of 0±24.5 counts.

- **Problem 18.11\*. Peak Fitting of Data.**

We are given an MCA data set consisting of a peak and constant background. We first try to visually estimate the centroid, area, and FWHM. The data in the list below corresponds to a channel number starting with channel 711, increasing by one, and ending with channel 728. Below we plot the data versus the channel number.

   **data = {238, 241, 219, 227, 242, 280, 409, 736, 1190, 1625, 1739, 1412, 901, 497, 308, 256, 219, 230}**

   **channel = {711, 712, 713, 714, 715, 716, 717, 718, 719, 720, 721, 722, 723, 724, 725, 726, 727, 728}**

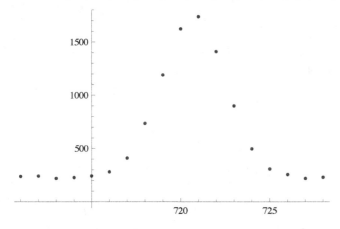

From the graph, we estimate a constant background of about 235 and FWHM of 4 channels, with peak channel at 721. We estimate the area by summing the data values and then subtracting an assumed constant background over the 5th through 17th data points:

   **area = (242 + 280 + 409 + 736 + 1190 + 1625 + 1739 + 1412 + 901 + 497 + 308 + 256 + 219) − (235 × 13)**

   **area = 6759**

(b). Now, we try the linearized method described in the text. We define $\ln[Q(x)] = \ln[\frac{y(x-1)}{y(x+1)}]$, do a linear least squares fit of $\ln[Q(x)]$ versus x over 5 channels, and plot this versus x. (In this problem the 5 channels we will use are channels 719 through 723) **We first subtract off the background** of 235 counts, and then proceed:

   **data = raw counts − 235**
   **= {3, 6, −16, −8, 7, 45, 174, 501, 955, 1390, 1504, 1177, 666, 262, 73, 21, −16, −5}**

We now plot x versus $\ln\left(\frac{\text{data}[\![i-1]\!]}{\text{data}[\![i+1]\!]}\right)$ where i is the element number of the data list, starting with 9 and ending at 13.

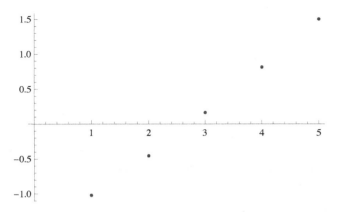

Now, let's do a linear least squares fit over data points (9 → 13). We'll skip the weighting factor (data[[i-1]] data[[i+1]]/(data[[i-1]]+data[[i+1]])) here. The (x, y) coordinates and the linear least squares fit line are shown below.

$$\begin{pmatrix} 719 & -\log\left(\frac{1390}{501}\right) \\ 720 & -\log\left(\frac{1504}{955}\right) \\ 721 & \log\left(\frac{1390}{1177}\right) \\ 722 & \log\left(\frac{752}{333}\right) \\ 723 & \log\left(\frac{1177}{262}\right) \end{pmatrix} \qquad y = 0.631443\,x - 455.069$$

We now find the centroid, FWHM and $\sigma$.

$$x_0 = \frac{-\text{intercept}}{\text{slope}} = -\frac{-455.069}{0.631443} = 721$$

$$\sigma = \sqrt{\frac{2}{\text{slope}}} = \sqrt{\frac{2}{0.631443}} = 1.78$$

Recognizing that the FWHM of a Gaussian is 2.35 time the standard deviation of the distribution:

$$\text{FWHM} = 2.35\,\sigma = 2.35\sqrt{\frac{2}{0.631443}} = 4.18 \text{ channels}$$

Now let's get the area, assuming no uncertainties in the centroid and $\sigma$:

$$\text{area} = \sqrt{2\pi}\,\sigma\, e^{\left(\frac{\sum_{i=9}^{13}\text{data}[i]\left(\frac{(\text{channel}[i]-x_0)^2}{2\sigma^2}+\ln(\text{data}[i])\right)}{\sum_{i=9}^{13}\text{data}[i]}\right)} = 6788$$

This is quite close to our original estimate of the net area.

*Extra*: Note that we could also apply a nonlinear fit to the data using the three fitting parameters $y_0$, $x_0$, and $\sigma$.

We find that a nonlinear fit for the data to the function $y(x) = y_0\, e^{\left(\frac{-(x-x_0)^2}{2\sigma^2}\right)}$ yields:

$$y_0 = 1518, \quad x_0 = 721 \quad \text{and} \quad \sigma = 1.78$$

and the net area is then:

$$\text{area} = \sigma \, y_0 \sqrt{2\pi} = 6774$$

The nonlinear fit is generally regarded as more accurate since it avoids the problem of a potential bias associated with biased estimators.

## Miscellaneous Detectors

- ### Problem 19.1.  Cherenkov emission by electron

(a).  We use the energy threshold relationship for Cherenkov radiation:

$$\text{Energy} = m_e\, c^2 \left( \sqrt{1 + \frac{1}{n^2 - 1}} - 1 \right) \qquad \text{where we substitute } n = 1.47$$

> **Energy = 0.186 MeV**

(b).  For this amount of energy to be given to a Compton electron, the minimum gamma ray energy corresponds to a 180° scattering angle, with $E_e = \left(2\,E_\gamma^2\right) / \left(m_e * c^2 + 2 * E_\gamma\right)$.  Solving this equation for $E_\gamma$ and substituting $E_e = 0.186173$ MeV gives us:

> $E_\gamma = -0.144$ **MeV** and $E_\gamma = 0.330$ **MeV**

Clearly, the negative energy is not physical, so our solution is 330 keV.

- ### Problem 19.3.  Liquid Xe attenuation at 30 keV.

Figure 6.18 tells us that 2 inches of gaseous Xe at 1 atm will attenuate ~20% of 30 keV x-rays.  But the density of Xe gas at STP is 5.85 g/liter while the density of liquid Xe is 3.52 kg/liter.  Recall:
$I(t)/I_0 = e^{-(\mu/\rho)\,(\rho t)} = 80$ % so the mass attenuation constant $\mu/\rho$ is

$$\mu_\rho = \frac{\mu}{\rho} = \frac{-\ln(.8)}{(\rho\, t)} \qquad \text{and when we substitute } \rho = 5.85 \text{ gram/Liter and } t = 5.08 \text{ cm}$$

$$\mu_\rho = \frac{0.0075087 \text{ Liter}}{\text{gram--cm}}$$

Using the previous expression, $t = \left(1/\rho\,\mu_\rho\right) \text{Log}[I_0/I(t)]$

$$t = \frac{1}{(\rho\,\mu_\rho)} \ln\!\left[\frac{1}{f}\right] \qquad \text{where we substitute } f = 0.5 \text{ and } \rho = 3.52 \text{ kg/Liter}$$

> $t = 262\ \mu\text{m}$

- ### Problem 19.5.  Electron traps:  Advantage or disadvantage?

Radiation interactions in a crystaline material produce electrons and holes.  If the application requires an immediate readout of the interaction, then the electron traps are undesirable because the desired signal is held up by the electron being trapped at the impurity site.  For most scintillators used in spectroscopy, this is a disadvantage.  However, for dosimeters, where one wishes to integrate the total number of events over a period of time, then trapping the elctrons at an impurity site is an advantage.  The electrons are later released under thermal or optical stimulation, and the resulting emission measures the integrated number of events.

## Problem 19.7. Activation and counting

(a). We need to find the induced activity for a 10 minute and 20 minute irradiation since the count rate is proportional to the induced activity. Since the induced activity is given by $A(t) = A_\infty(1 - e^{-\lambda t})$, the ratio we want is given by:

$$A = \frac{1 - e^{-\lambda t_1}}{1 - e^{-\lambda t_2}} \qquad \text{where we substitute } \lambda = \frac{\ln(2)}{2.3}, t_1 = 10 \text{ and } t_2 = 20 \text{ (all time units are in minutes)}$$

$$A = 0.953$$

This says that a 10 minute irradiation gives 95% of the counts of a 20 minute irradiation.

(b). Suppose we counted forever (instead of just 10 minutes). What would the gain be in the number of net counts? Recall that Counts $= k(1 - e^{-\lambda t_C})$ and assume the irradiation and wait time remain constant. So the factor increase in counts is given by:

$$\text{counts} = \frac{1}{1 - e^{-\lambda t}} \qquad \text{where we substitute } \lambda = \frac{\ln(2)}{2.3} \text{ and } t = 10$$

$$\text{counts} = 1.05$$

Thus, only a 5% increase in counts would be obtained by increasing the counting time from 10 minutes to infinity. This is because of the short 2.3 minute half life of this isotope -- it's mostly gone after 10 minutes.

## Background and Detector Shielding

- ## Problem 20.1. Background from $^{40}K$ in NaI

We note that the total number of NaI molecules in the detector will be $\rho$V*(Avogardro's Constant)/(Molecular Weight) where V= $\pi r^2 H$ for a cylinder. Natural potassium is 0.012% $^{40}K$, which has a half life of 1.26 x $10^9$ years. We are allowed only 1 cps in the detector volume. So the maximum number of K atoms is $\lambda$ (0.012%)K = 1 cps, or solving for K (where we are using the symbol K also for the number of potassium atoms):

$$K = \frac{(1\,/\,\text{second})}{0.012\,\%\left(\frac{\ln 2}{T_{1/2}}\right)} \quad \text{where we substitute } T_{1/2} = 1.26 \times 10^9 \text{ years}$$

$$K = 4.78 \times 10^{20} \text{ atoms}$$

But the total number of NaI molecules in the detector are:

$$\text{NaI} = \frac{\rho\,V\,(\text{Avogadro's Constant})}{\text{MolecularWeight}}$$

where we substitute $V = \pi r^2 H$, MolecularWeight = $\dfrac{150\,g}{\text{mole}}$, $\rho = \dfrac{3.67\,g}{cm^3}$, $r = \dfrac{7.62\,cm}{2}$ and $H = 7.62\,cm$

$$\text{NaI} = 5.12 \times 10^{24} \text{ molecules}$$

So the atom ppm of K that is allowable is (noting 2 atoms per NaI molecule):

$$\text{ppm} = \frac{K}{2\,\text{NaI}} \times 10^6 = 46.7 \text{ atoms of } K \text{ per million atoms}$$

- ## Problem 20.3. Cosmic ray pulse rejection

Cosmic rays tend to be high energy charged particles that have a low probability of interacting in the relatively low density gas in the proportional tube (i.e., low $\frac{dE}{dx}$). As a result, the pulses that they produce are small and can be rejected by a discriminator setting. In contrast, scintillators are solids and therefore their higher density (and often higher Z) leads to greater energy deposition and signal size.

- ## Problem 20.5. 2.22 MeV background line

Around concrete shields and other environmental sources that contain large amounts of hydrogen or $H_2O$, one often sees a 2.22 MeV background line in gamma spectrum. The source of this line is neutron capture on hydrogen, or $^1H\,(n,\gamma)\,^2H$, which has a Q value of 2.22 MeV. The neutrons are primarily produced by cosmic ray interactions in the atmosphere and local environment.